IFIP Advances in Information and Communication Technology 315

T0181155

IFIP – The International Federation for Information Processing

IFIP was founded in 1960 under the auspices of UNESCO, following the First World Computer Congress held in Paris the previous year. An umbrella organization for societies working in information processing, IFIP's aim is two-fold: to support information processing within its member countries and to encourage technology transfer to developing nations. As its mission statement clearly states,

> IFIP's mission is to be the leading, truly international, apolitical organization which encourages and assists in the development, exploitation and application of information technology for the benefit of all people.

IFIP is a non-profitmaking organization, run almost solely by 2500 volunteers. It operates through a number of technical committees, which organize events and publications. IFIP's events range from an international congress to local seminars, but the most important are:

- The IFIP World Computer Congress, held every second year;
- Open conferences;
- Working conferences.

The flagship event is the IFIP World Computer Congress, at which both invited and contributed papers are presented. Contributed papers are rigorously refereed and the rejection rate is high.

As with the Congress, participation in the open conferences is open to all and papers may be invited or submitted. Again, submitted papers are stringently refereed.

The working conferences are structured differently. They are usually run by a working group and attendance is small and by invitation only. Their purpose is to create an atmosphere conducive to innovation and development. Refereeing is less rigorous and papers are subjected to extensive group discussion.

Publications arising from IFIP events vary. The papers presented at the IFIP World Computer Congress and at open conferences are published as conference proceedings, while the results of the working conferences are often published as collections of selected and edited papers.

Any national society whose primary activity is in information may apply to become a full member of IFIP, although full membership is restricted to one society per country. Full members are entitled to vote at the annual General Assembly, National societies preferring a less committed involvement may apply for associate or corresponding membership. Associate members enjoy the same benefits as full members, but without voting rights. Corresponding members are not represented in IFIP bodies. Affiliated membership is open to non-national societies, and individual and honorary membership schemes are also offered.

Svetan Ratchev (Ed.)

Precision Assembly Technologies and Systems

5th IFIP WG 5.5
International Precision Assembly Seminar, IPAS 2010
Chamonix, France, February 14-17, 2010
Proceedings

 Springer

Volume Editor

Svetan Ratchev
The University of Nottingham
Precision Manufacturing Centre
NG7 2RD Nottingham, UK
E-mail: svetan.ratchev@nottingham.ac.uk

CR Subject Classification (1998): J.2, I.5, J.6, C.3, J.7

ISSN   1868-4238

ISBN-13   978-3-642-26241-8 Springer Berlin Heidelberg New York

springer.com

© IFIP International Federation for Information Processing 2010
Softcover reprint of the hardcover 1st edition 2010

Typesetting: Camera-ready by author, data conversion by Scientific Publishing Services, Chennai, India
Printed on acid-free paper   SPIN: 12844609   06/3180   5 4 3 2 1 0

Preface

The development of new-generation micro-manufacturing technologies and systems has revolutionised the way products are designed and manufactured today with a significant impact in a number of key industrial sectors. Micro-manufacturing technologies are often described as disruptive, enabling and interdisciplinary leading to the creation of whole new classes of products that were previously not feasible to manufacture. While key processes for volume manufacture of micro-parts such as machining and moulding are becoming mature technologies, micro-assembly remains a key challenge for the cost-effective manufacture of complex micro-products. The ability to manufacture customizable micro-products that can be delivered in variable volumes within relatively short timescales is very much dependent on the level of development of the micro-assembly processes, positioning, alignment and measurement techniques, gripping and feeding approaches and devices.

Micro-assembly has developed rapidly over the last few years and all the predictions are that it will remain a critical technology for high-value products in a number of key sectors such as healthcare, communications, defence and aerospace. The key challenge is to match the significant technological developments with a new generation of micro-products that will establish firmly micro-assembly as a mature manufacturing process.

The book includes the set of papers presented at the 5th International Precision Assembly Seminar IPAS 2010 held in Chamonix, France from the 14th to the 17th February 2010. The International Precision Assembly Seminar was established on 2003 by the European Thematic Network Assembly-Net to provide a forum for discussing the latest research, new innovative technologies and industrial applications in the area of precision (mini and micro) assembly.

The published works have been grouped into four parts. Part 1 is dedicated to micro-product design with specific emphasis on design for micro-assembly (DFμA) methods and solutions. Part 2 is focused on micro-assembly processes and includes contributions in process modelling, high-precision packaging and assembly techniques and specific examples of micro-assembly applications. Part 3 describes the latest developments in micro-gripping, micro-feeding and micro-metrology. Part 4 provides an overview of the recent developments in the design of micro-assembly production systems with specific emphasis on reconfigurable modular micro-assembly equipment solutions.

The seminar is sponsored by the International Federation of Information Processing (IFIP) WG5.5, the International Academy of Production Research (CIRP) and the European Factory Automation Committee (EFAC). The seminar is supported by a number of ongoing research initiatives and projects including the European sub-technology platform in Micro and Nano Manufacturing MINAM, the UK EPSRC Grand Challenge Project 3D Mintegration, the EU-funded coordinated action NanoCom and the EU-funded collaborative project FRAME.

The organisers should like to express their gratitude to the members of the International Advisory Committee for their support and guidance and to the authors of the papers for their original contributions. Our special thanks go to Luis Camarinha-Matos, Chair of the IFIP WG5.5, and Michael Hauschild, Chair of the STC A of CIRP, for their continuous support and encouragement. And finally our thanks go to Ruth Strickland and Rachel Watson from the Precision Manufacturing Centre at the University of Nottingham for handling the administrative aspects of the seminar, putting the proceedings together and managing the detailed liaison with the authors and the publishers.

February 2010   Svetan M. Ratchev

Organization

International Advisory Committee

T. Arai	University of Tokyo, Japan
H. Afsarmanesh	University of Amsterdam, The Netherlands
M. Björkman	Linköping Institute of Technology, Sweden
H. Bley	University of Saarland, Germany
C.R. Boer	ICIMSI-SUPSI, Switzerland
I. Boiadjiev	TU Sofia, Bulgaria
L.M. Camarinha Matos	University Nova, Portugal
D. Ceglarek	Warwick University, UK
A. Delchambre	ULB, Belgium
M. Desmulliez	Heriot-Watt University, UK
S. Dimov	University of Cardiff, UK
G. Dini	University of Pisa, Italy
S. Durante	DIAD, Italy
K. Ehmann	Northwestern University, USA
R. Fearing	University of California at Berkeley, USA
R.W. Grubbström	Linköping Institute of Technology, Sweden
C. Hanisch	Festo AG & Co, Germany
T. Hasegawa	Kyushu University, Japan
J. Heilala	VTT, Finland
J. Jacot	EPFL, Switzerland
M. Krieger	CSEM, Switzerland
S. Koelemeijer	Jaeger-Lecoultre, Switzerland
P. Lambert	ULB, Belgium
R. Leach	National Physical Laboratory, UK
N. Lohse	University of Nottingham, UK
P. Lutz	LAB, France
H. Maekawa	Nat. Inst. of Adv. Industrial Science and Technology, Japan
B. Nelson	ETH, Switzerland
J. Ni	University of Michigan, USA
D. Pham	Cardiff University, UK
M. Pillet	Polytech Savoie, France
G. Putnik	University of Minho, Portugal
B. Raucent	UCL, Belgium
K. Ridgway	Sheffield University, UK
G. Seliger	TU Berlin, Germany
W. Shen	Nat. Research Council, Canada
M. Tichem	TU Delft, The Netherlands
R. Tuokko	TUT, Finland
E. Westkämper	Fraunhofer IPA, Germany
D. Williams	Loughborough University, UK

Organization

International Advisory Committee

Table of Contents

Part I: Design of Micro Products

Chapter 1. Design for Micro-assembly

Chapter 2. Tolerancing for Micro-assembly

Part II: Micro-assembly Processes

Chapter 3. Development of Micro-joining Processes

Chapter 4. Innovative Assembly Processes

Chapter 5. Metrology and Control for Micro-assembly

Part III: Gripping and Feeding Solutions for Micro-assembly

Chapter 6. High Precision Positioning and Alignment Techniques

Chapter 7. Gripping and Handling Solutions

Part IV: Development of Micro-assembly Production Systems

Chapter 8. Modular Reconfigurable Assembly Systems

Chapter 9. Micro-Factory

Chapter 10. Micro-assembly Technology Studies

Part 1

Design of Micro Products

Part I

Design of Micro Products

Chapter 1

Design for Micro-assembly

Analysis of the Applicability of Design for Microassembly Theory to Biomedical Devices

Carsten Tietje[1], Daniel Smale[2], Steve Haley[2], and Svetan Ratchev[2]

[1] Division of Engineering, The University of Nottingham Ningbo, China,
199 Taikang East Road, Ningbo, 315100, China
carsten.tietje@nottingham.edu.cn
[2] Precision Manufacturing Centre, The University of Nottingham, University Park,
Nottingham, NG7 2RD, UK
daniel.smale@nottingham.ac.uk, svetan.ratchev@nottingham.ac.uk

Abstract. This paper describes the application of design for microassembly (DµFA) theory to the designing and assembling of biomedical microdevices in order to cope with the market-specific requirements of the biomedical sector which can be seen as one of the most complex industrial areas for microassembly applications. It is shown how DµFA can support the move from the research laboratory to industrial fabrication. The benefits of applying DFµA theory to the development of a biomedical microdevice are clearly shown, i.e. savings in cost and time achieved in the early design stages. Therefore a practical case study containing a minifluidics blood separation device is introduced providing insight into the process of guiding the design process of microproducts.

1 Introduction

The development and manufacture of biomedical[1] products for medical treatment represents a significant area in the European biomedical healthcare sector which is characterised by annual revenues in the order of 10 billion Euros [1, 2]. The development of biomedical devices is increasingly often accompanied by miniaturisation and functional integration and due to their intended area of application by complex environmental constraints.

The paper presented here describes the application of design for microassembly (DµFA) theory to the designing and assembling of such a biomedical microdevice in order to cope with the market-specific requirements. Furthermore, it is shown how DµFA can support the move from the research laboratory to industrial fabrication which is often hindered by problems in selecting and implementing appropriate micromanufacturing processes [3-5]. The key to tackling these issues can be seen in the design stages: at present there is no sufficient link between microproduct design and production system design.

---

[1] The term *biomedical* is refers to biotechnology-derived medical devices and products that are mainly acquired for the medical sector (EMCC, 2007).

S. Ratchev (Ed.): IPAS 2010, IFIP AICT 315, pp. 5–12, 2010.

The significant impact of the product design on the production and its cost has been analysed extensively [6-9]: it is easiest to make alterations early in the product development phases, consequently "[they are] the ideal and only time to get manufacturing cost right" [8].

2 Definition of the Microassembly Case Study

Assembly in the microdomain is more challenging than in the macrodomain because of different levels of maturity in the technology, differently occurring physical behaviour, and required microspecific processes. Although the differences have been elaborated on relatively extensively in the literature[2] there has been a lack in developing appropriate DFA theories for the microdomain [17]. The case study presented here aims at showing the benefits of applying DFµA theory to microassembly in the biomedical sector. In addition, it serves the purpose of verifying the methodology. The practical test case is introduced in section 2.1 before its assembly requirements are defined in section 2.2.

2.1 The Test Case – 3D Minifluidic Blood Separator

The biomedical sector can be seen as one of the most complex industrial areas for microassembly applications due to the arising of additional requirements such as cleanliness, high reliability, biocompatibility, tight tolerances, and governmental regulations. Manufacturing for the healthcare market is also characterised by requirements such as traceability and documentation (e.g. compliance to *Good Manufacturing Practices* or rules imposed by the *US Food and Drug Administration*).

The product introduced here to verify the applicability of DFµA theory is a minifluidics device consisting of several discs stacked on top of one another aiming at the separation of blood for analytic purposes. The product was developed as a demonstrator within the UK EPSRC grand challenge project *3D Mintegration* (3DM). The dimensions of the product (particularly the channels in the discs) are calculated and designed to enable blood flow and plasma separation.[3] Strong market relevance of the demonstrator is shown by fast growing rates of the healthcare sector worldwide, but particularly in the UK [18]. Therefore, establishment of cost-effectiveness is one of the main challenges imposing additional restrictions on the assembly process.

The joining process is one of the key processes in the manufacture of medical devices, this being due to the increasing sophistication of medical devices in terms of performance and therefore higher complexity of the devices' components. It is for this reason that the joining process has been singled out for this study as critical, imposing strict requirements (see section 2.2).

---

[2] On the differences between assembly in the macro- and in the microworld related to the required positional precision see [e.g. 10-13]. The changes in the physical behaviour such as scaling effects are analysed by [e.g. 14-16].

[3] Functioning is based on the flow behaviour of blood. The design is based on calculations and simulations carried out at the Universities Heriot-Watt, Greenwich, and Cambridge. The prototype parts were supplied by Cranfield University.

Consequently the application DFµA is focussed here on enabling the accurate joining of the parts demonstrating certain aspects of a methodology developed specifically for DFµA [19]. The following section described the parts and the joint requirements in more detail.

2.2 Requirements Definition

Five or more discs need to be joined on top of each other (in a stack). The discs, produced by micro-injection moulding, are 10mm in diameter and 1mm thick. The requirements with regard to the joining mechanism can be summarised as follows:

- *Accuracy of joint and placement*
  The parts need to be accurately aligned in order to allow device functioning. Perpendicularity of features to the first surface is required. Alignment accuracy of ±20 micrometres is needed in x- and y-directions, while rotational alignment is not necessary.
- *Hermetic seal*
  To avoid contamination of the blood sample a hermetic seal between the layers is necessary. This is also the precondition for realising the blood flow.
- *Contamination-free process*
  Contamination has to be avoided during the assembly. In addition, the parts' biocompatibility is not allowed to be effected.
- *Low cost*
  Low production costs are crucial in order to compete in the healthcare market, which is moving from analysis in expensive central laboratories to cost-effective point-of-care diagnostics.
- *High volumes*
  The device is envisioned to be disposable and therefore requires extremely high volumes, creating a need for very short cycle times.

The case study has been chosen according to the above requirements to illustrate that the methodology can have an impact on the mass market or when up-scaling production. This addresses problems currently existing when transferring research prototypes to industrial practice (see section 1).

3 Application of DFµA

Figure 1 shows the development process from the initial design idea to the parts to be assembled. The three-dimensional product has been conceptually designed within the 3DM consortium and an embodiment design and prototype parts have then been provided. Accordingly the DFµA methodology has been applied to existing parts. However, this offers advantages by allowing for testing the methodology's applicability and developing it further. As mentioned above, the key problem to be solved was how to join the parts while fulfilling the given requirements. The following subsections show how the process selection and the optimisation of the parts and joining process can be supported by the methodology.

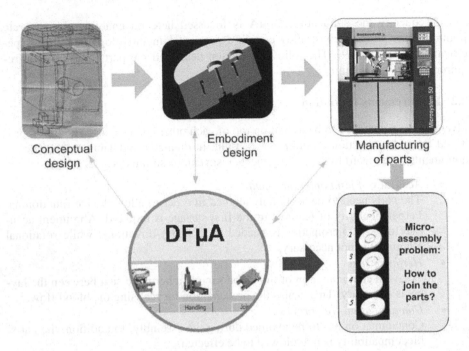

Fig. 1. Application of DFμA to the development process to the minifludics device

3.1 Process Selection

The assembly processes and equipment have been selected based on the defined requirements and the parts' characteristics (see section 2.2). Ultrasonic bonding was selected here particularly because of its cost-effectiveness and suitability to producing high volumes: the joining process can be automated to a very high degree with cycle times of below one second. In addition, ultrasonic bonding offers tight control with regard to dimensional tolerances while allowing for the realisation of very small joints on complex and fragile parts. Ultrasonic bonding can be optimised for use within a clean room environment, does not introduce contaminants or by-products, and does not interfere with the biocompatibility of the parts to be joined. Finally, the ultrasonic bonding process provides the required hermetic seal without subjecting the parts to high temperatures which would lead to thermal deformations. It should be noted that the ultrasonic bonding process can produce small particulate matter which may affect operation within a cleanroom environment. It would therefore be necessary to monitor the process and potentially fit mitigative equipment, such as pumps or screening.

Figure 2 shows a screenshot of the software prototype that was developed to support the process selection. It illustrates the description of the ultrasonic bonding process which can be seen as an ideal assembly approach for applications within the medical sector.

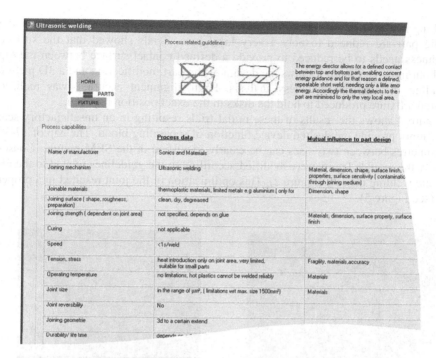

Fig. 2. Process sheet – Ultrasonic bonding (screenshot of the software prototype)

3.2 Optimisation of Parts and Joining Process

The optimisation and adaptation of the parts to be assembled and the selected joining process is a key step in the DFμA methodology. The assembly-oriented optimisation leads to parts that feature an area specifically designed to realising the joint (Figure 3). The parts contain an energy director that enables concentrated energy flow

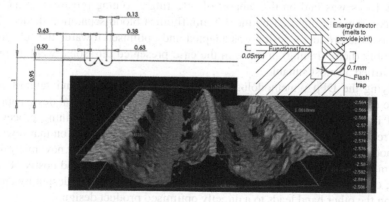

Fig. 3. Design adaptation for ultrasonic bonding - energy director

resulting in a defined and repeatable short weld. In that way, the thermal defects to the plastic part are reduced to only a very local area. Trials showed that the surface roughness needs to be reduced to provide a defined contact surface between the top and bottom parts. Figure 3 shows the design of the part modification and a 3D picture including the surface roughness analysis. High alignment accuracy was realised through a fixture produced to hold the disks in the exact position required.

Figure 4 shows the results of these initial trials resulting in an unsatisfactory seal that cannot provide the desired device function of separating blood. The gap of 17-30 micrometres between two discs can be clearly identified in the SEM picture. Consequently the moulded parts were modified according to the guidelines related to the ultrasonic welding process (Figure 2). This optimisation of the joint resulted in a proper seal (see Figure 5).

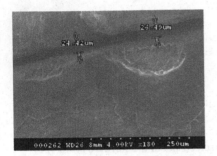 

Fig. 4. SEM analysis - no sealed joint   **Fig. 5.** Assembled device – hermetic joint

4   Conclusion and Outlook

The paper presented illustrates the benefits of the application of DFµA theory to the development of a biomedical microdevice. Therefore a practical case study was introduced providing insight into the process of guiding the design process of microproducts. A focus was laid on the support of selecting a joining process that meets the rigour requirements imposed by the 3D minifluidics blood-separation device. It was shown how the product's design was adapted and optimised towards the selected microassembly (here, joining) process. In the case presented ultrasonic welding process was identified as most suitable.

Applying the DFµA methodology shows that improvements and savings in cost and time can be achieved in the early design stages: the conventional design approach resulted in parts that were unusable, due to an unsatisfactory joining process, and needed reworking to fulfil the requirements of the test case. Based on that reworking, the embodiment design was changed, which resulted in the need for new micromoulds for the microinjection process. These steps are time-consuming and costly. The considering of process capabilities and related guidelines early in the design by applying DFµA on the other hand leads to a directly optimised product design.

Acknowledgments

The presented work is part of the UK EPSRC "Grand Challenge" research project "3D-Mintegration", that aims at *a paradigm shift in manufacturing by developing the technologies and strategic approaches required for the production of highly-integrated, cost-effective and reliable multi-functional 3D miniaturised/integrated devices.*

References

1. OECD, A framework for biotechnology statistics, Organisation for Economic Co-operation and Development (2005),
   http://www.oecd.org/dataoecd/5/48/34935605.pdf
2. EMCC, Trends and drivers of change in the biomedical healthcare sector in Europe: Mapping report. European Foundation for the Improvement of Living and Working Conditions European Monitoring Centre on Change (2007), http://www.eurofound.europa.eu/pubdocs/2007/28/en/1/ef0728en.pdf
3. Popovic, G., et al.: Examples for the technology selection method, in Micro fabrication processes - FSRM training course. FSRM, XIV-1 - XIV-8 (2004)
4. Alting, L., et al.: Micro Engineering. Annals of the CIRP 52(2), 635–657 (2003)
5. Hesselbach, J., Raatz, A. (eds.): mikroPRO, Untersuchung zum internationalen Stand der Mikroproduktionstechnik. Essen, Vulkan (2002)
6. Eversheim, W., Schuh, G. (eds.): Integrierte Produkt- und Prozessgestaltung. Springer, Berlin (2005)
7. Boothroyd, G., Dewhurst, P., Knight, W.: Product design for manufacture and assembly, 2nd edn. Marcel Dekker, New York (2002)
8. Miles, B., Swift, K.G.: Design for manufacture and assembly. Manufacturing Engineer 77(5), 221–224 (1998)
9. Reichenwald, R., Conrat, J.I.: Vermeidung von Aenderungskosten durch Integrationsmassnahmen im Entwicklungsbereich. Technical University Munich, Munich (1993)
10. Yang, G., Gaines, J.A., Nelson, B.J.: A flexible experimental workcell for efficient and reliable wafer-level 3D microassembly. In: IEEE International conference on robotics and automation, Seoul, pp. 133–138 (2001)
11. Scheller, T.: Untersuchung zu automatisierten Montageprozessen hybrider mikrooptischer Systeme, in Fakultaet Maschinenbau. Technical University Ilmenau, Illmenau (2001)
12. Cecil, J., Vasquez, D., Powell, D.: Assembly and manipulation of micro devices—A state of the art survey. Robotics and Computer-Integrated Manufacturing 23, 580–588 (2007)
13. Tichem, M., Lang, D., Karpuschewski, B.: A classification scheme for quantitative analysis of micro-grip principles. Assembly Automation 24(1), 88–93 (2004)
14. Van Brussel, H., et al.: Assembly of Microsystems. Annals of the CIRP 49(2), 451–472 (2000)
15. Fearing, R.S.: Survey of Sticking Effects for Micro-Parts. In: IEEE International Conference for Robotics and Intelligent Systems, IROS 1995, Pittsburgh (1995)
16. Ando, Y., Ogawa, H., Ishikawa, Y.: Estimation of attractive force between approached surfaces. In: Second Int. Symp. on Micro Machine and Human Science, Nagoya, Japan (1991)

17. Tietje, C., Ratchev, S.: Design for microassembly - capturing process characteristics. In: Dimov, S., Menz, W., Toshev, Y. (eds.) 4M2007 Conference on Multi-Material Micro Manufacture. CRC Press Borovets, Boca Raton (2007)
18. Ratchev, S., Hirani, H.: Synergetic process integration for efficient micro and nano manufacture - roadmapping stage 1 results. MicroSapient (2006),
http://www.microsapient.org
19. Tietje, C., Ratchev, S.: Design for Micro Assembly - A methodology for product design and process selection. In: IEEE International Symposium on Assembly and Manufacturing (ISAM). Omnipress, Ann Arbor (2007)

A Haptic Tele-operated System for Microassembly

P. Estevez[1], S. Khan[1], P. Lambert[1], M. Porta[1], I. Polat[2],
C. Scherer[2], M. Tichem[1], U. Staufer[1], H.H. Langen[1], and R. Munnig Schmidt[1]

[1] Precision and Microsystems Engineering, TU Delft
[2] Delft Center for Systems and Control, TU Delft
Mekelweg 2
2628 CD Delft
The Netherlands
Tel.: +31 (0) 15 2786428; Fax: +31 (0) 15 2782150
p.estevezcastillo@tudelft.nl

Abstract. A tele-haptic system for microassembly applications is currently being developed at the Delft University of Technology, with the goal of achieving superior performance by providing enhanced feedback to the human operator. Assembly of a micro-harmonic drive is used as a benchmark to fully evaluate the proposed tele-haptic system by investigating the control strategies and the individual subsystems: master device, microgrippers and slave system. The master device will be comprised of a parallel robot with a built-in gripper. The slave system and end effector are focused on providing efficient and effective force feedback of the interactions on the microenvironment to the human operator, in addition to detecting position and orientation of the object being grasped. Novel control strategies are also investigated to allow the transmission of high frequency transients to the operator, carrying information from hard contact interactions between the microgripper and the part to be assembled.

Keywords: Microassembly, Haptics, Teleoperation, Force Feedback, Parallel Robot, Event Based Haptics, Discretization.

1 Introduction

In today's emerging technologies, where the sizes of components are reaching the lower micrometer range, traditional automatic macro-assembly processes are being challenged to their limits. This is due to the requirements of high precision motion and small tolerances (usually less than a few micrometers). The predominance of surface forces over gravity complicates the release of parts, which tend to stick to the end-effector [1]. Furthermore, the handled parts are often delicate and fragile, requiring a control of forces in the micro-Newton range or below.

When dealing with microproducts that are produced in low-to-medium quantities with many variants, the automation of their assembly process may not be economically profitable because of the limited flexibility of the available assembly devices. For that reason the manual approach, often with the aid of a microscope, is

S. Ratchev (Ed.): IPAS 2010, IFIP AICT 315, pp. 13–20, 2010.
© IFIP International Federation for Information Processing 2010

presently the main method used in the assembly of low-to-medium series of microproducts. However, the pure manual approach is often not sufficient to fulfill the requirements, due to the difficulty of human operators to control the force and observe the precision aspects during the assembly.

Tele-operated systems [2], [3] are an interesting alternative to the unassisted manual assembly, but they fail on providing force feedback. Adding haptics to teleoperated systems becomes then a promising approach. The presence of the human operator, compared with the more rigid approach of automatic microassembly, improves the flexibility of the system thanks to his/her capability to plan, adapt, and react to unexpected situations during the assembly process. At the same time, the haptic-feedback reduces the risk of damaging handled parts and limits the assembly time, compared to regular tele-operated systems.

2 State of the Art

Tele-haptic systems have been used in different handling applications. Some researchers have utilized haptic interfaces in 1-DOF to perform basic operations like pushing [4] or more complicated functions for grasping micro-objects and carrying out assembly tasks [5], [6], [7]. Others have focused on biological applications by handling cells and tissue [8], [9]. Even if the results of using haptic technology for microassembly are promising, many challenges still need to be overcome with regard to design and control of multiple DOF haptic tele-operated systems for performing dexterous tasks. In particular there are different critical aspects that need research effort to improve the capabilities of the existing haptic environments:

   i)   The existing haptic environments present grasping force feedback, but no or little feedback in terms of *interaction force* (i.e. the force exerted on grasped microparts by other elements present in the assembly environment);
   ii) Obstacles and the trade-off between high magnification and depth of field, result in difficulties in the manipulation when only visual feedback is used;
   iii) Existing slave devices, such as [5] or [10], are not optimized for the presence of a human operator in the loop, resulting either in systems over-performing the user (and therefore unnecessarily expensive) or in non-transparent operations;
   iv) Current commercial haptic masters, such as the well known Phantom or the Omega devices, are not specifically designed for 2.5 D microassembly tasks.
   v)   Device specific control schemes have trade-offs leading to high deficiencies in some performance aspects, and the passivity based design techniques [11] still remain the mainstream control algorithms.

3 The Tele-haptic Environment and the Benchmark

The development of a tele-haptic microassembly system is currently in progress at the Delft University of Technology (TUD), within MICRONED program [12].

First, and in order to choose a particular benchmark application, an analysis was made of assembly applications containing typical and relevant micro assembly tasks.

This benchmark application is used to define the requirements for the master-slave robot combination and its tele-haptic environment. The analysis showed that many microassembly sequences consist of: pick-and-placing parts in relation to other parts, including defining a position in three translational degrees of freedom; aligning of parts around the vertical axis; and different kinds of peg-in-hole insertion, possibly requiring additional fine alignment on all 6 DOF. Some of the operations need fixing steps either by gluing, soldering, press fitting or releasing of flexible elements. Another important aspect pointed out by the analysis was that microproducts are often composed of fragile and delicate microparts and the risk of damaging them during the handling process is high. This is both due to high grasping forces, and environmental interaction forces on the manipulated parts. To avoid or at least reduce these risks, the control of the grasping and interaction force on the grasped microparts is of paramount importance.

A planetary and harmonic gearbox was chosen as benchmark application (Fig. 1) because its assembly process includes the typical assembly tasks described above. The assembly of this demonstrator–requires four DOF: three translations plus a rotation around the vertical z axis. The remaining rotations are expected to be less critical, and are planed to be addressed using the compliancy of the system and force sensing. Force feedback is required due to the fragility of the pegs, the low clearance, the slim gears and the need for part engaging. The parts of the demonstrator are going to be produced using MEMS technology, testing different shapes and dimensions.

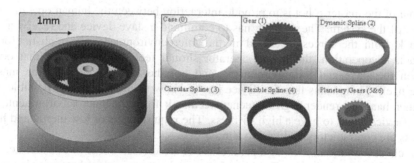

Fig. 1. The micro-harmonic drive benchmark and its components

As other tele-haptic systems, also this implementation is composed of a macro-side, the master, which interfaces with the operator, and a micro-side, slave, operating in the environment in which the assembly tasks are actually carried out. The two sides are in mutual communication by a suitable control system that allows inputs and outputs to be transferred between the macro- and micro-side. Fig. 2a shows the structure of the macro-side: a parallel robot with four DOF is used as human interface. A gripper handle provides the gripping interaction in terms of grasping motion and interaction force feedback. Fig. 2b shows the micro-side of the environment: a positioning robot with three translational DOF is provided with the support for the microparts, and a mechanical two-finger microgripper is supported by a 1 DOF rotational stage. Suitable devices have to sense the grasping and environmental interaction force on the grasped object, and two cameras focus on the microgripper to monitor the assembly operations. Other end-effectors such as different kind of grippers, a glue dispenser, laser microwelding and UV-light sources can substitute the two finger mechanical gripper or can be mounted on independent supports within the micro-side of the system. For the moment these extensions are not considered. The remainder of this section discusses the main components of the system independently.

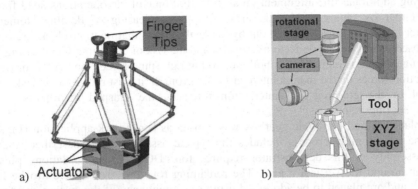

Fig. 2. The 4 DOF Master Robot and the 4 DOF Slave Robot with the gripper

3.1 Macro-side of the Environment

The role of the master robot is to provide interaction between the human operator and the rest of the system. The operator has to command a slave device and receive force feedback from the forces measured in the microenvironment. For realistic force-feedback teleoperation, the human operator should feel only the forces that occur between the slave and the environment, and not the forces present in the master device itself. This means that the device has to show as little inertia as possible. On the other hand, to render high-frequency feedback from the microenvironment, the master device needs to have a high stiffness. The requirements of low inertia and high stiffness are contradictory and lead to a trade-off.

The chosen solution relies on a parallel robot architecture. Due to this architecture usually a stiffness/inertia ratio is achieved that is about one order of magnitude higher than regular serial robots. An innovative parallel mechanism generating the four DOF motion has been designed (Fig. 2a). The master device is equipped with two finger tips that represent the mechanical gripper. The center point between the two tips can be moved in the three translations and the tips can be rotated around the center point. Finally, it is possible to open and close the tips around the center point to command gripping actions.

Various analyses concerning the useful workspace, the stiffness, and the dexterity of the mechanism have been done. Due to the complexity of the mechanical system, most of the analyses and optimizations have been carried out using interval analysis [13], a powerful method that reduces the computing time and allows us to guarantee the algorithm results. An optimization of the various link dimensions was carried out to maximize the useful workspace of the robot. The useful workspace is a centered spherical region without singularities and with acceptable robot dexterity. For a robot height of 200 mm, a preliminary optimization analysis gives us a 150 mm diameter sphere as useful workspace, where the gripping handle can be rotated ±65 degrees and the distance between the finger tips can vary from 40 to 80 mm.

3.2 Micro-side of the Environment

The slave system must position the end effector relative to the parts support following the commands given by the operator. A work space of 20 x 20 x 10 mm^3 must be covered, in order to reach all the micro-objects positioned on the microparts support, and sufficient resolution is needed to address assembly tasks with sub-micron clearances. The human actuation bandwidth of 3 to 10 Hz (depending on the kind of task) must also be matched. The requirements on drift and accuracy are less high, since the user is very capable of correcting for shortcomings in those aspects. Attention must be given to the possibility of mounting several end-effectors around the microassembly area, and to keep the area where the assembly is taking place always in focus. The presence of the human operator imposes additional constrains in order to obtain transparency and an intuitive operation.

The selected design, shown schematically in Fig. 2b, positions the parts support in three DOF, while the gripper rotates around the assembly point. By doing so, the vision system can stay focused on the microgripper/end effector, and the assembly area remains supervised regardless of the displacement of the microparts support. Furthermore, other tools can be mounted in the vicinity of the slave stage, and be used by positioning their end effectors into focus of the cameras. Adding a rotational degree of freedom to the base would improve the flexibility of the setup, but preliminary tests with a simulated environment suggest that it may be confusing for the human operator. Further psychophysical tests are been conducted to determine whether controlling all the degrees of freedom in the parts support, specially the rotational one, confuses the human operators affecting their performance, and to determine how to counteract this effect if this was the case. Additional psychophysical tests were conducted and consulted [14] to determine the average user capabilities for actuation and sensing, and map them into the slave workspace to ensure a transparent operation. Speeds of 400mm/s and accelerations of 4m/s were commonly observed in hand/wrist operation. Also, a range of 200 mm with resolution of 1 mm was identified (a 200:1 ratio), while the task covers a range of 20 mm with submicron resolution (a 100,000:1 ratio). Due to this difference, a multistep approximation to the goal will be necessary. In that way, for the roughest step a speed requisite of 40 mm/s and acceleration of 400 mm/s^2 is set, together with a resolution of 100 μm; the last and finest step must give a resolution of 40 nm with speeds of 16 μm/s and accelerations of 160 μm/s^2, over a workspace of 8x8x8μm^3. A parallel architecture is selected for the finest steps in order to minimize the moving mass, and flexure hinges are used to limit friction and provide high resolution motion.

The feedback architecture must provide two perpendicular views of the assembly area, force measurements from the gripping action, and force measurements from the interaction of the gripped part with the environment. Gripping forces should be high enough to hold, by friction, metallic and silicon based parts from tens to several hundreds of micrometers during transport and assembly, while avoiding damage to the slim gears and pegs. This yield to a requirement for sensing with a range up to 5mN and resolutions of micronewtons. Better resolutions would allow the user to control smaller interactions forces, protecting the parts and improving the positioning accuracy.

In order to provide the aforementioned visual feedback, two cameras will be located at a top and lateral position. For the environmental interaction force sensing two concepts are considered: inserting the force sensing device at wrist level (i.e. between the gripper and the support or between the base and the parts), or integrate interaction force sensors into the gripper itself. Solutions based on silicon microstructures with integrated sensors, strain gauges over metallic structures, or in measures of the positioning stage actuation power, are been compared in order to find the most appropriate principle. For the microgripper, silicon based grippers with integrated actuators and sensors are considered to be good solutions for satisfying the force requirements. The research has focused on the design of microgrippers with both grasping force and position sensing capability [15], [16]. The information about the object position within the gripper is usually detected by the visual feedback. The visual control however has intrinsic optical limitations such as its obstruction by obstacles on the workspace and the trade-off between depth of focus and magnification. Due to these visual control drawbacks, microgrippers that are able to detect the position and orientation of the object on the grasping surface can support and complement visual information, providing additional and important feedback. It can for instance detect whether the object is safely and correctly grasped or notice an unwanted movement of the object on the grasping surface. This information is of significant importance for successfully carrying out various assembly tasks [17]. Research efforts are still needed for providing these grippers with a suitable opening range to grasp micro objects of several sizes, up to a few hundreds of microns.

3.3 Control

The role of the controller is to match the position and force between the master and the slave system. 2-channels and 4-channels bilateral controllers have already been used in haptic teleoperation [18]. In haptic microassembly, the external forces acting on the work piece lead to position-dependent nonlinear characteristics. In order to avoid overly conservative control designs based on passivity techniques for ensuring stability and performances, it is of fundamental relevance to develop suitable structured uncertainty models (such as multivariable sector conditions on force nonlinearities) for the work-piece environment in haptic microassembly.

As a substantial benefit, these techniques even offer the opportunity to design controllers that adapt themselves to measurable changes of the environment in order to even further enhance performance. The current focus of the research is the hard contact stability problem, which is ubiquitous in the literature. The short-term goal is to characterize the stability properties of different control schemes, for instance [11], and develop a framework for robustness analysis specifically tailored to teleoperation.

Also, event-based haptic algorithms [19] are being researched because of their capability to present contact information to the user. It is for instance desirable for the user to acquire a rough idea about the stiffness of the material to be handled before performing any further procedures. Thus, research is being carried out to identify the material stiffness by means of tapping with the end-effector and generate corresponding signals for the operator.

4 Conclusion

In this paper the tele-haptic environment in development at the Delft University of Technology is proposed as an alternative to the pure manual micro-assembly. The system will constitute a flexible and intuitive system for the assembly of microproducts in small and medium quantities. The analysis of existing microproducts led to the definition of a micro-harmonic drive as a benchmark suitable for microassembly systems and processes. This analysis also showed that actuation on four DOF is required for this system, and that the presence of grasping and interaction force feedback would enhance the performance over a wide variety of microassembly tasks. The particular characteristics of the suggested task and the involvement of a human operator, motivate the specifications for the environment, defining a new and interesting niche where the system will continue its development.

Acknowledgment

The authors wish to thank the technical staff of PME and the DIMES IC Process Group at TU Delft for their precious help in the development of the devices. Special thanks are due to Dr. Jia Wei for his cooperation in the research on microgrippers. This research is supported by the BSIK MicroNed program [12].

References

1. Fearing, R.S.: Survey of sticking effects for micro parts handling. In: IEEE/RSJ Int. Workshop on Intelligent Robots & Systems (IROS 1995), Pittsburgh (1995)
2. Kunt, E.D.: Design and Realization of a Microassembly Workstation. MS Thesis, Sabanci University (2006)
3. Mitsuishi, M., Watanabe, T., Nakanishi, H., Hori, T., Watanabe, H., Kramer, B.: A tele-micro-surgery system across the Internet with a fixed viewpoint/operation-point. In: IROS 1995, Pittsburgh, Pennsylvania, USA (August 1995)
4. Khan, S., Sabanovic, A.: Force Feedback Pushing Scheme for Micromanipulation Applications. Journal of Micro-Nano Mechatronics (accepted, 2009)
5. Beyeler, F., Probst, M., Nelson, B.J.: A Microassembly System with Microfabricated Endeffectors for Automated Assembly Tasks. In: Workshop on Robotic Microassembly of 3D Hybrid MEMS, IEEE/RSJ Int. Conf. on Intelligent Robots and Systems (October 2007)
6. Zhou, Q., Aurelian, A., del Corral, C., Esteban, P.J., Kallio, P., Chang, B., Koivo, H.N.: Microassembly station with controlled environment. In: Proc. of SPIE, vol. 4568, pp. 252–260 (2001)
7. Hériban, D., Agnus, J., Gauthier, M.: Micro-manipulation of silicate micro-sized particles for biological applications. In: 5th Int. Workshop on Microfactories, Besançon, France (2006)
8. Sieber, A., Valdastri, P., Houston, K., Eder, C., Tonet, O., Menciassi, A., Dario, P.: A novel haptic platform for real time bilateral biomanipulation with a MEMS sensor for triaxial force feedback. Sensors and Actuators A 142, 19–27 (2008)

9. Arai, F., Sugiyama, T., Luangjarmekorn, P., Kawaji, A., Fukuda, T., Itoigawa, K., Maeda, A.: 3D viewpoint selection and bilateral control forbio-micromanipulation. In: IEEE Int. Conf. on Robotics and Automation (ICRA 2000), vol. 1, pp. 947–952 (2000)
10. Perroud, S., Codourey, A., Mussard, Y.: PocketDelta: a miniature robot for microassembly. In: 5th International Workshop on MicroFactories, Besançon, France (October 2006)
11. Colgate, J.E., Schenkel, G.G.: Passivity of a Class of Sampled-Data Systems: Application to Haptic Interfaces. Journal of Robotic Systems 14(1), 37–47 (1997)
12. Microned Programme, http://www.microned.nl/
13. Moore, R.E., Bierbaum, F.: Methods and Applications of Interval Analysis. Society for Industrial Mathematics (1979)
14. Tan, H.Z., Srinivasan, M.A., Eberman, B., Cheng, B.: Human factors for the design of force-reflecting haptic interfaces. In: 3rd Int. Symposium on Haptic Interfaces for Virtual Environment and Teleoperator Systems, Chicago, IL, vol. 55/1, pp. 353–359 (1994)
15. Porta, M., Wei, J., Tichem, M., Sarro, P.M., Staufer, U.: Vertical Contact Position Detection and Grasping Force Monitoring for Micro-Gripper Applications. In: IEEE Sensors 2009 (2009)
16. Wei, J., Porta, M., Tichem, M., Sarro, P.M.: A Contact Position Detection and Interaction Force Monitoring Sensor for Micro-Assembly Applications. In: Transducers 2009, Denver, CO, USA, pp. 2385–2388 (2009)
17. Porta, M., Tichem, M.: Grasping and Interaction Force Feed-Back in Microassembly. In: Accepted to the 5th Int. Prec. Assembly Seminar (IPAS 2010), Chamonix, France, February 14-17 (2010)
18. Hannaford, B.: A design framework for teleoperators with kinesthetic feedback. IEEE Transactions on Robotics and Automation 5(4), 426–434 (1989)
19. Kuchenbecker, K.J., Fiene, J., Niemeyer, G.: Improving Contact Realism Through Event-Based Haptic Feedback. IEEE Transactions on Visualization and Computer Graphics 2(2), 219–230 (2006)

Neutral Interface for Assembly and Manufacturing Related Knowledge Exchange in Heterogeneous Design Environment

Minna Lanz, Roberto Rodriguez, Pasi Luostarinen, and Reijo Tuokko

Tampere University of Technology, Department of Production Engineering,
P.O. Box 589, 33101 Tampere, Finland
{minna.lanz,roberto.rodriguez,pasi.luostarinen,
reijo.tuokko}@tut.fi
http://www.tut.fi/tte

Abstract. The goals of this research are to provide overview of the recent activities in the field of design of manufacturing and assembly processes from the knowledge share point of view and propose a solution how to connect products to the manufacturing or assembly processes and suitable systems. The research conducted here starts with the assumption of connectivity of three design domains: product, process and system design domains. The characteristics of the product are pre-describing the set of the processes needed to manufacture and/or assemble the product. The description of the product defines constraints for the suitable processes. The processes are pre-defining the system requirements and constraining the set of systems capable of carrying out the needed processes. From the technical point of view the goal is also to provide an information architecture, where the relations exist between these three domains.

Keywords: Ontology, knowledge exchange, knowledge base.

1 Introduction

Global economy has made manufacturing processes to become more and more distributed around the globe. Design teams are located possibly in every continent. In such a geographically and temporally divided environment, effective and proficient collaboration between design teams is crucial to maintain product quality, production efficiency and organizational competency.

Partly due the global the nature of manufacturing facilities has been in continuous change in past decades. The emergence of highly computerized design and control systems are changing the fundamental assumptions of how the production should be planned and controlled. Centralized systems based upon the need to share data by point of need acquisition are giving a way to global distributed production network enabling as well as requiring localized and fast adaptation to the production changes.

In the recent years industrial standards are being defined in a more computer readable form, most notably since the emergence of XML-based formats and computing power. XML as a language has a number of advantages for developers and implementers,

S. Ratchev (Ed.): IPAS 2010, IFIP AICT 315, pp. 21–29, 2010.

because these specifications can be compiled by computers, databases can be automatically built, and certain kinds of testing can be performed more easily. However, some groups have used XML markups as a substitute for modeling the information - a dangerous shortcut that only works in communities that already share a common understanding of the meaning and usage of terms. Based on Ray [8] a far better approach is to adopt one of the emerging semantic technologies, such as Web Ontology Language (OWL), or first order logic.

2   Challenges from the Industry

There are several barriers for achieving an integrated collaborative production facility. One of these is the challenge with the proprietary information representations, which often force the experts to serve as manual human-machine interfaces between otherwise automatable functions with the concomitant introduction of errors, ambiguity and misinterpretation. This leaves the systems unable to share the information with required speed and accuracy. Today the problem is somewhat solved, unfortunately with considerable expense coming from highly customized information mapping, translations and add-hoc implementations knit together forming unstable structure of mainly shallow (meaningless) information [8], [2].

This leads to several problems; firstly the communication between departments' "designer domains" becomes time consuming since the models have to be created over and over again.

The second problem is that by every remake and update the model actually loses information, because the second tier does not require all of the information created in the first phase. The result is that there are multiple sources contributing specific knowledge to several isolated models and revisions instead into "the master model" [2].

The third problem that comes next these two is the re-use of the knowledge. At the moment the re-use of the existing information is quite impossible due to the several models which are meaningless from the knowledge-point of view. For example the production planning knows that there are several stations, robots and grippers which can be re-used in the production of the next product family needing only a bit of modification. However, despite the amount of stored information, there are no up-to-date complete information set easily available of the interfaces and life-cycle-data and dimensions of the station and its components. In most of the cases the new station and/or the line is therefore designed starting with an empty layout [2].

The fourth problem which follows, is that the even the semiautomatic reasoning becomes impossible. The decision support systems are at best unreliable and relying heavily on the users' expert knowledge of the product, processes and resources. There is a lot of potential for the re-use of the existing knowledge of the models. But once again the problem is that there is no definition, the ontology, which would explain what the system is, what its capabilities are and how long it has operated or why it was designed as it was in the first place. And even if the ontology would exist, the modern CAD/CAPP systems which are currently the market leaders cannot utilize the existing information [2].

The aim of the paper is to introduce a model for combining design information from several different sources under one reference model which can be accessed via common interface. The model itself utilizes feature-based modeling methodology as a core in order to combine product, process and system related information to the closed models of existing design tools.

For the product knowledge representation there exist the well known product knowledge sharing formats such as Standard for the Exchange of Product Model Data (STEP), [6], and its extensions. For process and system descriptions there are also different standards addressing the needs from those domains, such as Process Specification Language (PSL) and Core Manufacturing Simulation Data (CMSD). However, these standards as well they are defined can be seen as "islands of standards", since there are no models for representing all of these domains under one architecture [8]. In most of the implementations there are overlaps between standardized models but again they are not interconnected. The second note has to be made, that the standards have evolved over the years and become very complex set of extensions, dedicated to one view point and serving mainly that specified view point.

There exist different implementations of standards linked to each other, but very often these are forming a framework for expert systems, where the actual reasoning and design rationale is embedded into the model itself. The specified, yet detailed models are in most of the cases incomplete to be used as such in the case scenarios. The purpose of this work is to form a demonstration of the model where the different domains are indeed connected under one model in such manner that the model is general enough to be used in different case scenarios of which one is introduced in the Chapter Case.

3 Approach

3.1 Theoretical Approach

Currently the knowledge is exchanged between two or more systems through dedicated interfaces. When the company's design tool platform includes up to hundred different design tools and software the point-to-point integration between all of the systems becomes quite impossible. In order to manage the information exchange between systems, the exchanged information is stripped and only the basic geometry or simplified data structures are exchanged. This leads to the situation where the design is done in strongly filtered snap-shot information. In order to address the process and system related information through product description and reason based on the existing information another approach must be adapted.

The right side of figure 1 illustrates other option for information exchange by utilizing common information model which is accessed through one interface. In this case all of the overlapping design systems can contribute and retrieve their knowledge into a common knowledge model. In this case the focus is in contribution of knowledge rather than filtering and re-creating it. The clients here naturally cannot utilize fully all of the information but can use what they need and consider the remaining information as metadata. Metadata can be accessed and interpreted by humans as well.

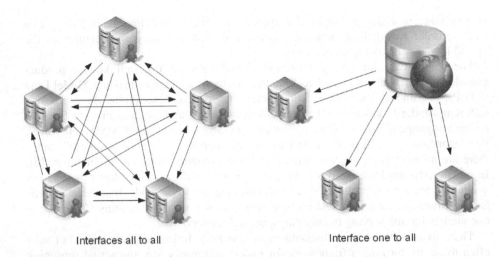

Interfaces all to all   Interface one to all

Fig. 1. Knowledge exchange between clients

The approach utilizes the theoretical methodology of product, process and assembly system connectivity graph introduced by Rampersad [7], who focused on the theoretical model of how the connectivity was shown in between of these three domains. Lohse [4] continued developing the integrated assembly model by concentrating to the connectivity of assembly processes and available assembly systems. The approach taken here concentrates the connection of products and processes as the starting point.

3.2 Implementation

The domain ontology for representing the connections, Core Ontology, illustrated in figure 2, was created with Protege 3.3.1 Owl editor and visualized with GraphViz. The Core Ontology consists of three domains for describing a content and context of a model. The product section of the ontology has four main levels *product, sub-assembly, part* and *feature*. The connection between products and processes is done in such way that the *Processes* are connected to the *Product level, Task to Sub-Assembly, Operations* to *Part*, and *Actions* to *Geometric Feature* level. The resource model consists of classes for devices, actors, tools, and area (having subclasses for factory, line and station) for describing the context of the manufacturing environment. The model allows the occurrence of multiple devices such as robot and drill combinations.

The Knowledge Base (KB) was designed to be a system where the data could be stored and retrieved for and by different applications varying from the product and process design to simulation. The first step in the development was to design an architecture that will allow the contribution and sharing of information between different applications. Figure 3 shows the planned architecture for the knowledge base. To achieve this design the following tools were used to facilitate the approach: Apache Tomcat web server, Jena semantic web framework [1], Pellet reasoner [5], and Postgre database

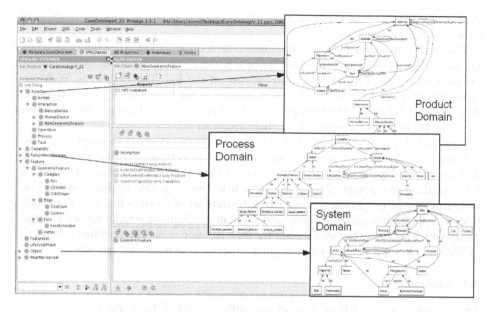

Fig. 2. CoreOntology for addressing Product, Process and System knowledge

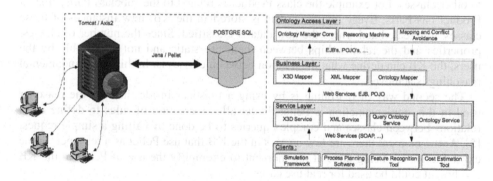

Fig. 2. Structure of the KB architecture and its implementation

One way to fulfill the challenges discussed in the second chapter is to use a standardized way to communicate with the clients. For that a Service Oriented Architecture (SOA) was chosen to be used in this implementation. SOA is defined by W3C [12] "*A set of components which can be invoked, and whose interface descriptions can be published and discovered*". SOA is a standardized architecture that can be implemented by using a set of Web Services, which are software systems designed to support interoperable machine-to-machine interaction over a network and an interface described in a machine-processable format e.g. WSDL (Web Service Description Language). Other systems interact with the Web service in a manner prescribed by its description using SOAP-messages, typically conveyed using HTTP with an XML serialization in conjunction with other Web-related standards [10]. With the use of

Web services the client applications could interact with the system by sending or retrieving information for the KB.

Originally, the architecture was designed to host different kind of services, depending on the needs of the applications. As result, the client applications were, indeed, the ones who pushed forward the development of the KB. The first set of Web services was developed based on the classes defined in the Core Ontology, see figure 2. After a while, when the development of applications started, additional Web Services were needed to fulfill the needs of the client applications.

The Web Services are segmented into five categories in order to simplify the development of the KB, but also to group Web services with similar functionality. This division is only logical, since all of them run on the same Web server and was done to improve the development of the code, since it is unmanageable to maintain and correct errors on the programming code when the number of services has grown. In addition, it simplifies the tasks of developers of client applications since it allows to target easily which services they would need for their development. The resulting categories of Web services are the following: product, process, system, model and general.

3.3 Reasoning

The reasoning in the KB is done in two different ways. First, the Core Ontology stored in the KB is founded in classes. Each of the classes has properties and relations to other classes. For example the class Product is related to the subclass Parts, since a Product has Parts. Every time new data is stored in the KB, new instances of those classes are created and where the new data is classified. Since the number of classes, properties and the relationships between them are static and not modifiable by the users, the KB can define what the relation between classes is. In this way fundamental reasoning is achieved.

The second way of reasoning is by using a reasoner inside the KB, the reasoner used in this implementation is Pellet. Pellet allows creation of rules to define extra relations between classes and complex queries to be done by calling a simple request [5]. At the moment there are few queries in the KB that use Pellet as a reasoner. Some extra queries are being done at the moment to exemplify the use of Pellet in the KB and how it could be used for real use cases.

4 Case Study

The case study is inspecting and testing the developed ontologies as knowledge exchange media between the manufacturing and assembly design systems in heterogeneous environment, while all of the clients are utilizing proprietary data structures. In this case the process plan of the product is first created in the process planning tool, CAMeLEAN. The planning of manufacturing/assembly processes with CAMeLEAN is dedicated to the processes and resource description in the shop floor. CAMeLEAN describes the processes in a workstation and focuses for balancing processes between the work-stations of a complete production line. The processes generate or manipulate parts of the final product or the product itself. Processes have cycle-times, predecessors and successors and depend on resources defined via this tool [11].

The goal of the case study is to inspecting the aspects of how to link product and process models and to use those (semi)automatically in the simulation tool. The underlying idea is to utilize the ontological product and process definition to populate the simulation environment as automatically as possible. Currently the client, illustrated in the figure 4 built inside the Visual Component's 3DCreate simulation environment access straight to the KB client and queries whether or not the product description has reference parts or visual components connected to the ontology. The process plan is simulated inside the 3DCreate and validated processing times are updated to the KB. Moreover, in case that changes there have been made on the layout or resources used the corrected models can be uploaded to the KB.

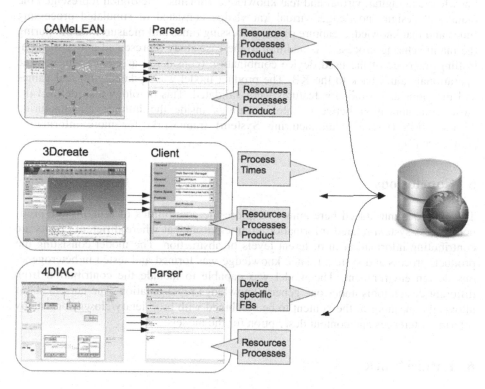

Fig. 4. Required knowledge exchange between the design tools

The Framework for Distributed Industrial Automation and Control (4DIAC) is being integrated with the PiSA SP3 tool chain for programming the various control devices within the production process. The control programs developed with the 4DIAC-IDE (Development Environment) are programmed with the IEC 61499 standard and uses Function Blocks as the main building blocks [9]. Within the PISA project, the 4DIAC-IDE is connected to the KB and stores Function Block Types (FBT) within the KB, as can be seen in the figure 4.

The interaction between CAMeLEAN and 4DIAC takes places via the KB. CAMeLEAN defines the preliminary process layouts and resource planning including the hardware components being used within the process. 4DIAC retrieves the processing times and resource information and adds the detailed representation of hardware elements and corresponding FPs to the KB.

The case study explored the defined integrated ontology from the knowledge contribution and share point of view. The clients here are used in more in the design phase than actual manufacturing. The clients are able to modify, add or delete one or all of the instances concerning the particular product or process description.

Another case study was conducted where the same ontology and knowledge architecture was used for conveying the planned processes and corresponding product to the manufacturing environment. The Core Ontology was able to capture the design knowledge in digital, virtual and real knowledge domains. The digital represented the plans and design knowledge, virtual knowledge consisted of simulated processing times and real knowledge capture of the processing times and measured values during the manufacturing process. As addition to the case study represented above the controlling interface of the used device combination (considered as client) returned the operational values back to the KB. The product model was updated with real values and the simulation times for features were validated. This ontology also defined the basic communication between holonic entities inside the holonic manufacturing TUT's DiMS (Digital Manufacturing System) framework, for more information please see [3].

5 Conclusions

The approach introduced here aims to explore the possibilities of utilizing the product-process-system related information in an environment where different clients are contributing information in different levels of abstraction. The model consisting the product, process and system related knowledge was formed and tested in heterogeneous design environment. The model was capable to include the contribution from different design tools under one generic knowledge representation. The current status allows the meaning of the content to be combined with proprietary/closed formats by offering references and content description for the models.

6 Future Work

The CoreOntology used at the moment is very generic by nature and the future work focuses on developing more detailed resource models to capture capabilities, geometric properties and other metadata relevant to that particular resource. The ontology needs to be improved in such manner that it can capture and convey the operational parameters of each resource into the simulation environment where the different production scenarios can be tested before applying into the factory floor. Further more, the ontology must be able to capture both the history of what happened to the product and the temporal description of each resource. By saving the resource's operational history, the backtracking of the systems' conditions, energy consumptions or parts' design features becomes more feasible. In addition, the tool's current use state and other machinery's characteristics can be saved to other KB and used for improving the process efficiency, pinpointing the possible quality problems related to the part's features or single machines and overall improvement of various processes.

Acknowledgments. The authors would like to thank 6th Framework Programme Integrated Project PISA - Flexible Assembly Systems through Workplace-Sharing and Time-Sharing Human-machine Cooperation consortium of their valuable feedback during the design and implementation of the system.

References

1. Jena, http://jena.sourceforge.net/
2. Lanz, M., Kallela, T., Velez, G., Tuokko, R.: Product, Process and System Ontologies and Knowledge Base For Managing Knowledge Between Different Clients. In: IEEE SMC International Conference Distributed Human-Machine Systems, pp. 508–513 (2008)
3. Lanz, M., Tuokko, R.: Generic Reference Architecture for Digital, Virtual and Real presentations of manufacturing systems. In: Proceedings of Indo-US Workshop on Designing Sustainable Products, Services and Manufacturing Systems (2009),
   http://www.mel.nist.gov/publications/publications.cgi
4. Lohse, N.: Towards an Ontology Framework for The Integrated Design of Modular Assembly Systems. PhD thesis, University of Nottingham (2006)
5. Pellet, http://www.mindswap.org/2003/pellet/
6. Rachuri, S., Han, Y.-U., Foufou, S., Feng, S.C., Roy, U., Wang, F., Sriram, R.D., Lyons, K.W.: A Model for Capturing Product Assembly Information. Journal of Computing and Information Science in Engineering 6(11) (March 2006)
7. Rampersad, H.K.: Integrated and Simultaneous Design for Robotic Assembly (1994) ISBN- 10: 0471954667
8. Ray, S.R.: Tackling the Semantic Interoperability of Modern Manufacturing Systems. In: Proceedings of the Second Semantic Technologies for eGov Conference (2004)
9. Rodriguez, R., Lanz, M., Rooker, M.: Task 3.4.1. Implementation of the Knowledge Base. PiSA SP3 Task report, p. 22 (2009)
10. Thomas, E.: Service-Oriented Architecture: A Field Guide to Integrating XML and Web Services. Prentice Hall, Upper Saddle River (2004)
11. Velez Osuna, R.: SP3 Task 3.1.4 Conceptual Toolset Definition. PiSA SP3 task report 3.1.4, p. 38 (2007)
12. World Wide Web Consortium, Service-Oriented Definition W3 Glossary,
    http://www.w3.org/2003/glossary/keyword/All/
    ?keywords=service-oriented

Chapter 2

Tolerancing for Micro-assembly

Defining Tolerances in Assembly Process with the Aid of Simulation

H.-A. Crostack[1], R. Refflinghaus[1,*], and Jirapha Liangsiri[2]

[1] Dortmund University of Technology, Chair of Quality Engineering
Joseph-von-Fraunhofer Str. 20, 44227, Dortmund Germany
[2] Carlsberg Breweries A/S,
Ny Carlsberg Vej 100, DK-1760 Copenhagen, Denmark
robert.refflinghaus@lqw.mb.tu-dortmund.de

Abstract. Analysis and allocation of tolerance are the typical problems in tolerance planning for the assembly and its components. This paper presents the developed simulator which is able to assist in solving tolerance planning problems in production processes, both in manufacturing and assembly processes. The paper also shows the benefits of this simulator which leads to quality characterisation and improvement, cost reduction and shorter design and planning phases.

Keywords: Tolerancing for assembly, Tolerance planning, Assembly process, Simulation.

1 Introduction

Nowadays it is unlikely that any producer or manufacturer will produce their products without any assembly process. The more usual case is that a producer will have several components manufactured by themselves or suppliers and then assembled together. With the introduction of tolerances, it becomes possible to join these components together by the assembly process, since the specification of required tolerances assures that the components will fall within certain cut-off dimensions and fit together. The requirements from the engineering and manufacturing sides are not in the parallel direction. Engineers like tight tolerances to assure fit and function of their designs, however, these can result in excessive process costs. Manufacturers prefer loose tolerances which make parts easier and less expensive to produce. But these may lead to increased waste and assembly problem. Therefore, tolerances must be planned carefully in order to provide a common meeting ground where competing requirements can be solved. The task of tolerance planning is the finding of the optimum where the product can fulfil customer's requirement with optimum cost and time [6].

Planning tolerances for a finished product consists of many components created by different processes is not an easy task since the quality of assemblies depends on the quality of the manufactured components to be assembled. When the components do not fall within the tolerance, the difficulties can occur in one of the two following

---

* Corresponding author.

S. Ratchev (Ed.): IPAS 2010, IFIP AICT 315, pp. 33–40, 2010.

ways. Firstly, the intended components cannot be assembled together. Secondly, even though the intended components can be assembled together, the assembly process may require a longer processing time than expected. This would result in the higher production cost. Therefore, the planner has to determine which combination of component tolerances is the best. Simulation is chosen in this research as an approach to evaluate the tolerance planning for products which are produced particularly by manufacturing and assembly processes. The evaluation aims to improve overall quality, cost, and time in the production.

The paper explains the basic background and the concept of the developed simulator. Before ending the paper with the conclusion, the validation result of the simulator is presented.

2 Tolerance Planning

There are two main types of problem in tolerance planning; tolerance analysis and tolerance allocation. In tolerance analysis, the component tolerances are all known or specified and the resulting assembly variation is calculated by summing the component tolerances to determine the assembly variation. In tolerance allocation the assembly tolerance is known from design requirements, whereas the magnitudes of the component tolerances to meet these requirements are unknown. The available assembly tolerance must be distributed or allocated among the components in some rational ways [1].

Simulation can help in tolerance analysis by, for example, simulating the effects of manufacturing variations on assemblies to obtain the assembly function. The assembly tolerance can be determined from this assembly function based on the required yield or acceptance fraction.

For tolerance allocation, the component tolerances can be adjusted until the desired assembly's quality is acquired. Many researchers such as Wu [7] and Jeang [4] used simulation with other algorithms such as design of experiment to determine the optimal tolerance design in an assembly.

This section has shown that simulation can be used as a helping tool in tolerance planning and tolerancing problem solving. The simulator, which is the outcome of this research, can help as well in tolerance planning and tolerancing problem solving. It provides the opportunity to analyse both individual component or part and the whole assembly by taken quality, cost, and time into account. In order to achieve the most benefit in production, tolerance adjustment should be done together with implementing a good inspection strategy. With this simulator, the tolerance planning can be done at the same time as inspection planning.

3 Simulator for Tolerance and Inspection Planning

The simulator was developed to be able to model each production process and to simulate the quality characteristics which are produced by each process. A mathematical model was developed and integrated into the simulator in order to include the effect of manufacturing process on assembly process.

The new developed simulator focuses on the tolerance and inspection planning and their influences on the production processes. It is designed to investigate the impact of different component's tolerances and inspection strategies on overall production cost, cycle time, and product quality. The flow chart of SixSigma components is shown in Figure 1. SixSigma consists of three components, which are Manufacturing, Assembly, and Inspection process.

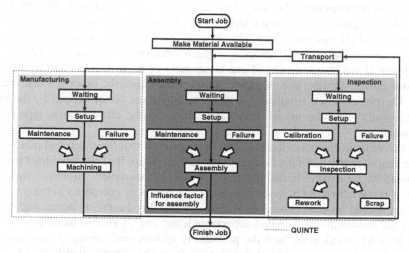

Fig. 1. Components of SixSigma simulator, showing as well the processes derived from QUINTE simulator

The concept of the manufacturing and inspection processes are taken from another simulator called QUINTE and adapted into SixSigma simulator. QUINTE was built to simulate the machining and inspection in a detailed way. It aimed to investigate the impact of different inspection strategies on manufacturing cost, cycle time, and product quality [2].

3.1 Manufacturing Process

Initially, the model of machining process is characterized by its own statistical distribution depending on the current process capability. Then, from this model, SixSigma randomly simulates the quality characteristic value of the given process. SixSigma models the distributions dynamically since the process capability is not constant over time. This phenomenon was described by Nürnberg [6]. The expected value can glide from its original value or the deviation can increase because of failure, wear of the tool, etc. This changed distribution can be restored or improved by setting up and maintenance.

From this dynamic model, the simulator generates a random number belonging to the specified distribution at the time of simulation. The obtained characteristic value denotes the actual value of a manufactured part's characteristic.

3.2 Assembly Process

The quality characteristics from assembly process can also be presented with statistical distributions. The accuracy of most assemblies is determined primarily by the accuracies of the individual components to be assembled. Therefore, these effects of the components' quality which have on the quality of the assemblies are necessary to be included in the simulator.

The simulator simulates the characteristic of the assembly in two different cases. The first case applies if the characteristics of all components are conforming. The characteristic of the assembly are randomly generated by simply using the distribution which represents the assembly process in the normal process condition. The second case applies if the characteristic of one component or more is non-conforming. The simulator cannot simulate the characteristic by using either the process's distribution alone. The model which simulates the characteristic must be influenced by the component's quality and the model differs depending on the characteristic type of the component.

For the variable characteristic case, the assumption is that the machined part, which has the characteristic that falls outside the specification limit of the machined part but still inside the acceptable limits, can still be possibly assembled. However, the probability of non-conforming in assembly process increases as compared to the probability in the normal process condition. And the probability keeps on increasing until it reaches the minimum or maximum acceptable limit. Therefore, distance ratio is introduced in order to calculate how the probability of non-conforming characteristic in assembly process changes. The distance ratio from the normal distribution is illustrated in Figure 2 [3].

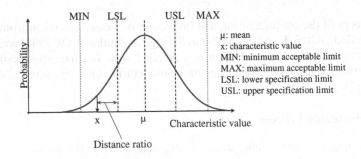

Fig. 2. Example of distance ratio

In the current simulator, it is assumed that the probability of non-conforming in assembly process increases linearly. The ratio of distance of machined characteristic can be calculated by the following equations:

When X is lower than lower specification limit,

Distance ratio$_j$ = $(LSL_j - X_j) / (LSL_j - MIN_j)$

When X is higher than upper specification limit,

Distance ratio$_j$ = $(X_j - USL_j) / (MAX_j - USL_j)$

Where   MIN is minimum acceptable limit
            MAX is maximum acceptable limit
            LSL is lower specification limit
            USL is upper specification limit
            j is characteristic j; for $j = 1,...,m$

The distance ratio gives no effect on the assembly's quality when the component's characteristic falls inside the specification limit. Thus it is set to zero. The assembly part turns out to be a bad part when the distance ratio falls out of the acceptable limit or the distance ratio is set to be equal to one.

Each machined characteristic has different impact on the quality of the assembly. Another variable called importance factor is initiated at this point to integrate these impacts into the model. The importance factor, which can be determined from the historical data, ranges from zero to one. The multiplication of distance ratio and importance factor of characteristic is called the characteristic's influence factor.

The influence factor of the component is the combination of influence factors of related characteristics. Furthermore, the influence factor of the whole assembly can be found from its related components' influence factor. It is assumed that the effects of component characteristics on the assembly are independent from each other, thus the influence factor for component and assembly can be derived from the following formulas:

$$IF_i = 1 - \prod_{j=1}^{m} [1 - (DR_j * ImpF_j)]$$

$$IF \text{ for assembly} = 1 - \prod_{i=1}^{n} [1 - IF_i]$$

Where   IF is influence factor
         DR is distance ratio
         ImpF is importance factor
         i is component i; for $i = 1,...,n$
         j is characteristic j; for $j = 1,...,m$

The influence factor for assembly will affect the model on assembly's process distribution. Besides, the distribution is influenced in time by failure and maintenance as in manufacturing process. Therefore, the assembly can obtain the characteristic value from the new adjusted model.

3.3   Inspection Process

The inspection process is simulated in the similar way as manufacturing process. Due to bias and precision, the value given by the inspection tool may differ from the true value. The capability of the inspection process is described by a statistical distribution, for example, a normal distribution. A standard deviation $\sigma insp$ and a mean $\mu insp$ are assigned for each inspection process. The machined characteristic value is used as a mean for the inspection process. The expected value can glide from its original value or the deviation can increase due to failure. And the distribution can be restored or improved by setting up and calibration.

SixSigma randomly generates inspected value from the specified distribution. The inspected value will be compared with the specification limit, thus deciding whether the part is conforming or not.

After the decision is made, the part, which is declared as a conforming part, continues on its production sequence. Scraps must be sorted out and rework parts can be handled in two ways. The rework parts can be sent back to the preceding process or processes and repeat the operation. Another option is to repair the part at separated rework area.

4 Validation and Implementation

The simulator was validated with both fictional and actual cases [5]. A preliminary validation of the simulator was done with a fictional case to validate the use of simulation in evaluating different tolerances and inspection strategies The result from this preliminary validation, which is shown in Figure 3, is compared with the logical trend.

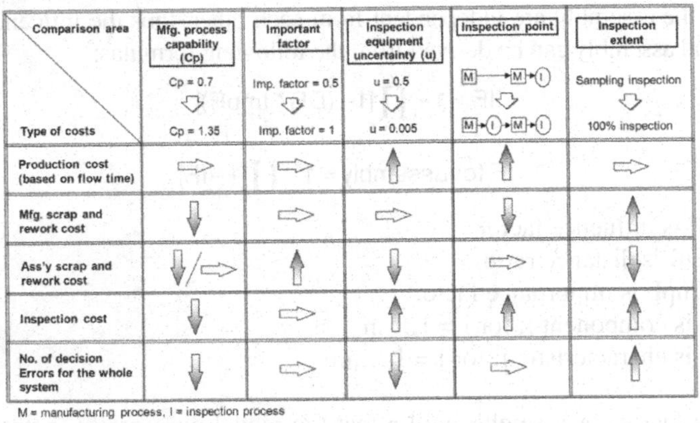

Comparison area / Type of costs	Mfg. process capability (Cp)	Important factor	Inspection equipment uncertainty (u)	Inspection point	Inspection extent
	Cp = 0.7 ⇩ Cp = 1.35	Imp. factor = 0.5 ⇩ Imp. factor = 1	u = 0.5 ⇩ u = 0.005	M →M·I ⇩ M·I→M·I	Sampling inspection ⇩ 100% inspection
Production cost (based on flow time)	⇒	⇒	⇑	⇑	⇒
Mfg. scrap and rework cost	⇓	⇒	⇒	⇓	⇑
Ass'y scrap and rework cost	⇓/⇒	⇑	⇓	⇒	⇓
Inspection cost	⇓	⇒	⇑	⇑	⇑
No. of decision Errors for the whole system	⇓	⇒	⇑	⇒	⇑

M = manufacturing process, I = inspection process

Fig. 3. Simulation results for preliminary validation

The comparisons of the results with the logical trend were done in order to investigate if the simulator is able to simulate the effects of manufacturing process on assembly process or not and how they effect on the assembly process. Additionally, comparisons of different inspection strategy aspects with respect to quality, cost, and time were conducted to investigate how the production performance fluctuates according to the changes in different strategy. The results prove that the model, which assigns an influence factor for each assembly part according to the quality and importance of its manufactured components, was successfully developed and validated.

A more extensive validation was done by implementing the simulator at a pilot company. The simulator was used to identify the optimal tolerance planning and inspection planning of one final product which consists of several components. The actual historical data was used as an input for the simulation. One of the validations was done particularly on the simulator's function in tolerance planning. This validation was experimented on a final product with 2 components (component A and B). The tested strategies were differentiated by the components' tolerances. The result in Figure 4 shows that only the tolerance of component B significantly affects the quality of the finished product. From this investigation, the company can know which component's tolerance they should tighten or improved. In this way, they will be able to properly plan the tolerances of components and achieve the require quality level of the finished product. They can also make the break-even analysis between the investments that they have to make versus the degree of improvement that they can gain.

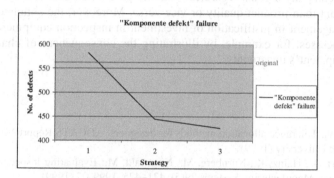

Fig. 4. Comparison on number of defects of different tolerance planning strategies

Moreover, the scenarios with different strategy for tolerance planning and inspection planning were created and ran with the simulation. The simulation gave the result which suggested several improvement points for the company. The company gained the insight of their processes and realized which process they should pay attention to. The output of the simulation were broke down into several types of costs, defects, and time used at each process. Figure 5 shows an example of the simulation result which was obtained from the pilot company case.

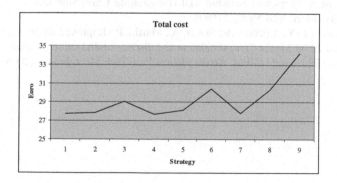

Fig. 5. One of the simulation results of pilot company

5 Conclusion

The validation proved that the simulator can assist in investigation and evaluation of the tolerance and inspection planning, so that the suitable tolerances and inspection strategy can be selected without the risk of trying it out in the real production environment. The output from the investigation leads to quality improvement, cost and time reduction. Usually the three aims are contradictory because a higher product quality causes higher costs and requires more time. Thus, the weights for the single goals must be set by each company individually. In this way, the simulator can help to improve the company's goals regardless whether production cost and cycle time are more important than product quality or vice versa. Moreover, the simulator can support the management in justification of investment in inspection equipment or manufacturing processes, for example, by illustrating the consequences of changes in inspection equipment's uncertainty.

References

1. Chase, K.W.: Tolerance allocation methods for designers, ADCATS Report No. 99-6. Brigham Young University (1999)
2. Crostack, H.-A., Heinz, K., Nürnberg, M., Nusswald, M.: Evaluating inspection strategies by simulation. Manufacturing Systems 29(5), 421–425, 1999-07 (1999)
3. Crostack, H.-A., Heinz, K., Mayer, M., Höfling, M., Liangsiri, J.: Simulation im Qualitätswesen – Der Einsatz des Planungstools QUINTE+ zur Optimierung der Prüfplanung, Tagungsband. In: ASIM Fachtagung Simulation in Produktion und Logistik, Berlin, Germany, October 4-5 (2004)
4. Jeang, A.: Optimal tolerance design by response surface methodology. International Journal of Production Research 37(14), 3275–3288 (1999)
5. Liangsiri, J.: Assembly Process Improvement by Means of Inspection Planning and Corresponding Tolerance Planning - A Modelling and Simulation Approach, Dissertation, Universität Dortmund (2007)
6. Nürnberg, M.: Ein Beitrag zur Entwicklung der Toleranzplanung auf der Basis von Risikobetrachtungen. Fortschritt-Berichte VDI (Dissertation Universität Dortmund), Reihe 20, Nr. 292, Düsseldorf: VDI Verlag (1999)
7. Wu, F., Dantan, J.-Y., Etienne, A., Siadat, A., Martin, P.: Improved algorithm for tolerance allocation based on Monte Carlo simulation and discrete optimization. Wuhan University of Technology, China, and LGIPM, Arts et Métiers ParisTech Metz, France (2008)

Assembly Analysis of Interference Fits in Elastic Materials

Kannan Subramanian and Edward P. Morse

Structural Integrity Associates, Inc., 11515 Vanstory Drive,
Suite 125, Huntersville, NC 28078
ksubramanian@structint.com
Center for Precision Metrology, Dept. of ME&ES,
University of North Carolina at Charlotte,
Charlotte, NC 28223
emorse@uncc.edu

Abstract. The objective of this work is to provide fast approximations of force/work/energy calculations for designers of assemblies in which interference is required. In this paper, the subjects of interest are assemblies in which there are several sets of features that can influence the final location of the parts, once assembled. The analysis includes a systematic approach in developing an analytical model involving basic laws of equilibrium and more advanced finite element based models to estimate the work required for assembly and the total strain energy in the components after assembly is complete. In this approach, an ideal press-fit type interference assembly is considered initially and solution methodology is developed. A non-symmetric hinge assembly with multiple interferences is analyzed later with the developed approach and the results are compared with the experimental observations. The suitability of the GapSpace assembly analysis method [1] for assemblies in which there is interference between the components is also investigated.

1 Introduction

Due to the increasing performance requirements, geometric tolerances on machine components are being driven smaller and smaller. In order to avoid excessive 'slop' or free play between components, it may be desirable to have a small, but non-zero, interference between the components. Press-fit is commonly chosen for faster and cheaper assembly of such an interference-type fit, instead of the cumbersome shrink-fit. These assemblies contain several sets of features that can influence the final location of the parts, once assembled. The interference assembly undergoes contact pressure and friction forces at the contact interface which is a common characteristic of any contact problem. Analytical solutions are difficult to find for even simple geometries due to the non-linear conditions resulting from frictional forces at the contact surfaces. With the advent of high performance computers and efficient computational techniques such as finite element method, complicated interference geometries can also be analyzed with a fair degree of accuracy. In this work, the interference problem

S. Ratchev (Ed.): IPAS 2010, IFIP AICT 315, pp. 41–49, 2010.

is considered in a broader and general sense by analyzing a simple interference fit and developing approaches for complex interference assemblies. The key feature of this work is to investigate the assembly force, work done, and strain energy of the press-fit assemblies. A commercially available finite element analysis package, ANSYS 11.0 [2], has been used for analyzing the interference assemblies considered in this work. The ultimate objective is to provide tools to assist designers with fast approximations of the parameters (work, force, and energy) mentioned above.

2 Simple Problem

An idealized symmetrical interference assembly involving three members is considered (see Figure 1) for the theoretical development and validation purposes. Assuming that the interference is very small, that is, $\delta \ll c$, all the materials in the assembly are the same and follow linear elastic and isotropic properties, and in the absence of any friction, a simplistic analysis from solid mechanics principles described in [3], results in the following displacements for the three members in the assembly,

$$\text{change in length of member 1 } (\delta_a) = \delta \frac{a}{L},$$

$$\text{change in length of member 2 } (\delta_b) = \delta \frac{b}{L},$$

$$\text{and change in length of member 3 } (\delta_c) = \delta \frac{c}{L}, \text{ where } L = a + b + c.$$

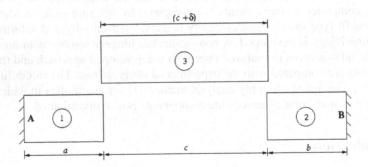

Fig. 1. Idealized interference fitting problem

Due to the symmetry associated with the model, only half of the geometry shown in Figure 2 is modeled in ANSYS with rounded mating corners to overcome numerical instabilities associated with sharp corners. Member 3 is assumed to move incrementally into the space between members 1 and 2. The contacting surfaces were assumed to be flexible-to-flexible type and 8-noded linear quadrilateral elements are used to mesh the geometry. Plane strain conditions are assumed. The surfaces in contact are assumed to be sliding freely without friction. For a geometry with a = b = 50 mm, c = 100 mm with an interference of δ = 1 mm, the displacement results for each member using finite element analysis is found to be in good agreement with the results calculated with the force equilibrium based approach discussed in Reference 3.

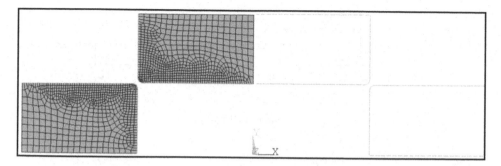

Fig. 2. Finite Element Model for the Simple Problem

That is, a displacement of 0.25 mm at the end of the assembly is calculated ($\delta_a = \delta_c/2 = 0.25$ mm) irrespective of the material under consideration. In this analysis, low alloy carbon steel material is considered with an Elastic Modulus of 200 GPa and yield strength of 1600 MPa. The ANSYS model included a dry friction co-efficient of 0.2. Displacement contours at the end of the assembly are shown in Figure 3, and it can be observed that the member 1 and 2 are under compression with a maximum displacement of approximately 0.25 mm which is consistent with the analytical results.

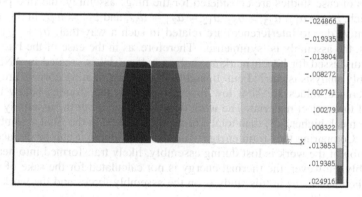

Fig. 3. Displacement (in mm) Results for the Simple Problem

2 Hinge Assembly

With the preliminary bench marking described in the previous section, the assembly analysis is extended to a 'hinge assembly' for various combinations of fitting conditions. Figure 4(a) represents such a generic and parametric hinge assembly, Figure 4(b) represents the four GFS pairs in the assembly, and lastly Figure 4(c) characterizes the fitting conditions, and the inequalities needed for non-interfering assembly. The identification of these fits conditions for non-interfering assemblies is described in Reference 1. The extension to this method which is the subject of the current work is simply to relax the requirement that all fits conditions have non-negative values as shown in Figure 4(c).

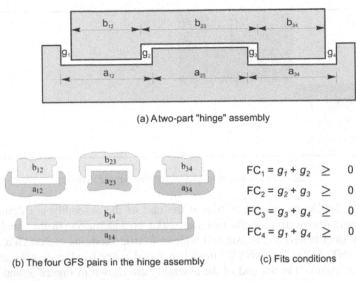

(a) A two-part "hinge" assembly

(b) The four GFS pairs in the hinge assembly

(c) Fits conditions

$$FC_1 = g_1 + g_2 \geq 0$$
$$FC_2 = g_2 + g_3 \geq 0$$
$$FC_3 = g_3 + g_4 \geq 0$$
$$FC_4 = g_1 + g_4 \geq 0$$

Fig. 4. General Hinge Assembly with Various Fits Conditions

A series of case studies are considered for the hinge assembly, the first press-fit is one in which, $b_{12} > a_{12}$, $a_{23} > b_{23}$, $a_{12} = a_{23} = a_{34}$, and $b_{12} = a_{34}$. In this case, the gaps (negative due to interference) are related in such a way that, $g_1 + g_2 = g_3 + g_4$, that is, the assembly is 'symmetric.' Therefore, as in the case of the basic simple geometry discussed for bench marking, only a half model is created in ANSYS and the assembly analysis is carried out. In addition, the sharp mating corners are rounded and interference values are kept low in order to keep the stresses below the yield strength of the subject material. The work done in carrying out the assembly is calculated to be much higher (x8) than total strain energy stored in the final assembled configuration. Comparing the strain energy with the work done, it can be deduced that a large fraction of the work is lost during assembly, likely transformed into heat during the assembly. However, the thermal energy is not calculated for the sake of simplicity. Also, the main focus in this study is on the assembly forces and the work required for such an assembly.

The second and third of the case studies are carried out for non-symmetric cases with differing interferences and are characterized by, $b_{12} > a_{12}$, $a_{23} > b_{23}$, $b_{34} > a_{34}$, $a_{12} = a_{23} = a_{34}$, and $b_{12} > b_{34}$. The gaps are selected in such a way that, $|g_1 + g_2| > |g_3 + g_4|$. The total interference in the third case is double that of the second case. Due to the non-symmetry, the whole geometry is modeled in ANSYS and analyzed. In these cases, the calculated stresses are higher in the region of greater interference in line with the physics of the problem (see Figure 5c). The last two cases considered are those in which one feature pair of the assembly "fits perfectly" while other sections have interference. These "just-fit" cases are characterized by $b_{12} > a_{12}$, and $a_{12} = a_{23} = a_{34}$. In addition, for the first just-fit case, $a_{23} = b_{23}$, and $b_{34} > a_{34}$, while for the second just-fit case, $a_{23} > b_{23}$, and $b_{34} = a_{34}$. The gaps are selected in such a way that, $|g_1 + g_2| > |g_3 + g_4|$. For the second just-fit case, $g_3 = g_4 = 0$. Graphically, in the first of the two just-fit cases, the center sections of the parts are of the same undeformed size, while in the second case, the rightmost sections are of the same dimension. The stresses and displacement patterns correlate well with the physics of the assembly process. That is, more stresses are observed at the interfering sections while negligible stresses and zero displacements are observed at the just-fit sections.

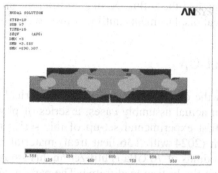

(a) Symmetric Hinge Assembly

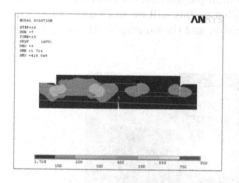

(b) Non-symmetric Hinge Assembly - 1

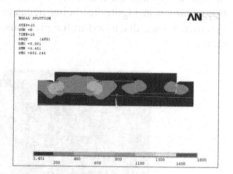

(c) Non-symmetric Hinge Assembly - 2

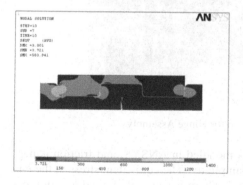

(d) Just-Fit Type Hinge Assembly - 1

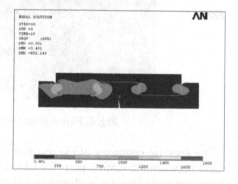

(e) Just-Fit Type Hinge Assembly - 2

Fig. 5. Equivalent Stresses (MPa) in various simulated Hinge Assemblies

The equivalent or von-Mises stress distribution in the assembled configuration for various hinge assemblies are shown in Figure 5. The boundary conditions imposed for the ANSYS analysis include that the lower part is restrained in the vertical direction and a few nodes at the bottom surface of the specimen are fixed to avoid rigid body motion in the horizontal direction. The top surface of the top member is free to move horizontally (to accommodate the influence of the different interference conditions), and given uniform vertical displacements until it comes to the final assembly position.

3   Experimental Set-Up

In an effort to establish the correlation between the numerical results and the forces and stresses observed in actual assembly cases, a series of physical experiments are being developed. The first experimental set-up of this series is shown in Figure 6. High strength aluminum (2024 with T36 heat treat) material, with yield strength of 325 MPa and Elastic Modulus of 70 GPa is utilized in this study. The aluminum parts are pressed together using the Instron load frame. The parts in assembly are machined in such a way that the assembly will contain multiple interferences with all fits conditions (shown in Figure 2c) being negative. That is, this will be an example of a non-symmetric case discussed in the previous section and shown graphically in Figure 5b and 5c.

Fig. 6. Experimental Set-up for Hinge Assembly

The specific geometry of the test parts is modeled in ANSYS, and friction is neglected between the mating parts for added simplicity. The equivalent stress at the end of the assembly process is shown in Figure 7. There is some plastic deformation occurring at the filleted regions, and hence the stresses are very high in those areas. However, in the overall bulk geometry of the components, the stresses are well below the yield strength of the material under consideration.

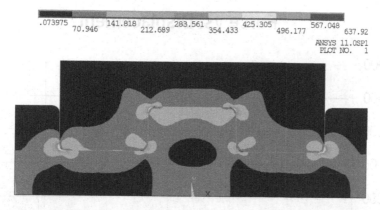

.073975   141.818   283.561   425.305   567.048
       70.946   212.689   354.433   496.177   637.92

ANSYS 11.0SP1
PLOT NO.   1

Fig. 7. Equivalent Stress Distribution at the Assembly Condition for Experimental Assembly

4   Results and Discussion

The assembly forces were calculated from ANSYS by selecting the top surface and integrating the vertical stress components over the length of the part for each deformation step. Figure 8 shows the comparison of these forces with the Instron-measured compressive forces. The red squares in the graph represent the calculated forces using ANSYS and the solid line represents forces measured during assembly.

During the initial contact, ANSYS predicted very high assembly force value while the measured forces are still lower. This is due to the modeling of the filleted region, and the adjustment of the parts as they become aligned for assembly. The assembly forces reduce as the analysis progresses in ANSYS and eventually begin following the pattern of the experimental observation – a steady increase as more of the parts overlap. The measured forces peaks between 8 and 10 kN, while the calculated forces reach 16kN, thus differing by a factor of approximately 1.6. The calculation of the work done during the assembly process is 38.6 Joules based on the measured forces, while the ANSYS calculated work is 104 Joules. The strain energy stored in the assembly is calculated to be 198 Joules. We believe this result is anomalous, and due to the manner in which ANSYS calculates strain energy. The strain energy is higher because the calculation of strain energy includes the elements that are in contact where plastic deformation may occur. However, the work done was calculated only based on the assembly forces and displacements.

The difference in results between the measured and ANSYS calculated work/force/energy can be attributed to several factors, such as contact friction, initial assembly deformation, and how accurately the ANSYS model represented the experimental set-up. The results presented here, while preliminary, are encouraging in that the magnitude of the forces and energy required for assembly are simulated with some agreement. A series of further experiments will help in developing a unified methodology that will predict the assembly forces.

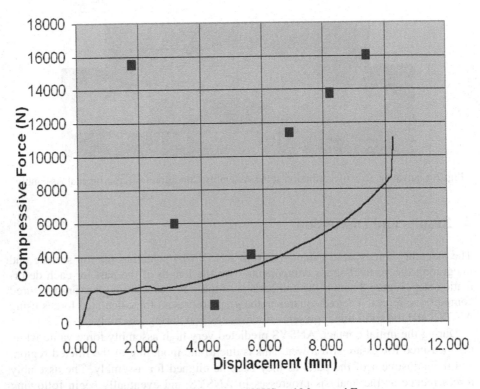

Fig. 8. Comparison between ANSYS and Measured Forces

5 Conclusions

It is demonstrated that the final location and displacement of the members of an assembly can be predicted using the identification of fitting conditions and relative starting position of the members. The simplistic cases allows to easily predict the outcome of the analyses, and the continuation of this work will utilize additional numerical experiments to predict the part displacement and location of the members based on the fitting conditions determined using GapSpace techniques. By having a set of test cases that span the variety of interactions possible in the assembly, guidance can be provided to the designer that will allow him or her to more quickly define nominal interference conditions that satisfy design requirements for force, stress, or assembly work. It is clear from the analyses that the interaction of all four of the fits conditions is important to the stress distribution in the components. Additional work will also be required to assess the sensitivity of these design objectives to the dimensions actually obtained in assembly. This will assist in the specification of geometric tolerances on the assembly features, and will be an important next step in assisting the designer.

Acknowledgments

The authors acknowledge the support of The University of North Carolina at Charlotte's Center for Precision Metrology, the William States Lee College of Engineering, and the National Science Foundation for its support of initial parts of this work through grant DMI-0237501. Any opinions, findings and conclusions or recommendations expressed in this material are those of the authors and do not necessarily reflect the views of the National Science Foundation. The assistance of Prof. Harish Cherukuri in developing the analytical results is also appreciated.

References

[1] Zou, Z., Morse, E.P.: A gap-based approach to capture fitting conditions for mechanical assembly. Computer-Aided Design 36(8), 691–700 (2004)
[2] ANSYS/Mechanical, Release 11.0 (w/Service Pack 1), ANSYS Inc. (August 2007)
[3] Subramanian, K., Morse, E.P.: Analysis of interference fits in elastic materials. In: Proceedings of DETC 2005, ASME 2005 Design Engineering Technical Conferences and Computers and Information in Engineering Conference, Long Beach, CA, September 25-28 (2005)

How Form Errors Impact on 2D Precision Assembly with Clearance?

Pierre-Antoine Adragna[1], Serge Samper[2], and Hugues Favreliere[2]

[1] LASMIS, Université de Technologie de Troyes, 12 rue Marie Curie, 10010 Troyes, France
[2] SYMME, Université de Savoie, BP 80439, 74944 Annecy le Vieux Cedex, France
adragna@utt.fr, {serge.samper,hugues.favreliere}@univ-savoie.fr

Abstract. Most models of assembling simulations consider that form errors are negligible, but how can this assumption be assessed? When clearances are high, form deviations can be neglected, but on the case of very precise mechanisms with small clearances, this assumption can lead to non-accurate models. This paper is the continuation of our previous works presented at IPAS 2008 dealing with the assembly of two parts regarding their form deviation. The proposed method considers the positioning of the pair of surface with a given external force to identify contact points. The parts relative positioning is expressed by a small displacement torsor that can be transferred to any referee and compared to the functional requirement. The objective of this paper is to identify the clearance domain of a mechanical linkage regarding the form deviation of parts. Several parameters are identified as influent such as the clearance value, the straightness of the form deviation and the localization of the ideal least squared associated shape.

Keywords: Form deviation, relative positioning, clearance domain, small displacement torsor, modal parameterization.

1 Introduction

Tolerancing of assembly can be solved using mathematical model of different levels of complexity. One very simple model presented by Graves [1] that only considers dimensional variations. More complex models are proposed to exploit the 3D tolerance zone, such as Chase [2] who proposes the use of the vector chain, or Davidson [3] who proposes the T-maps® model that can be compared to the small displacement torsor (SDT) proposed by Bourdet [4] and also used by Giordano [5] into the clearance domain model. These previous works consider that form deviation of parts can be neglected. Ameta [6] proposes to study the influence of form deviation based on the T-maps® model. Radouani [7] presents an experimental study of the positioning of parts regarding form deviations. Some mathematical methods are also presented to determine the relative positioning of parts regarding form deviations such as Neville [8], Stoll [9] and Morière [10]. Most of these approaches use optimization algorithm considering the minimization of a criterion based on distances or volumes between parts.

S. Ratchev (Ed.): IPAS 2010, IFIP AICT 315, pp. 50–59, 2010.

Adragna [11] and [12] presents the static method that simply calculate the relative positioning of one shape on another considering their form deviations and a contact force, leading to the resolution of a mechanical static equilibrium. The main idea is to transfer the form deviation of the first shape to the second one in order to have a similar but simpler problem. Due to the form deviation, only few points of the shapes can be in contact. The identification of all possible set of contact points is obtained by the computation of a convex shape that filters and identifies the possible contact points and facets. The consideration of a positioning force allows identifying the contact facet. The positioning problem is then simplified to the identification of a geometric transformation in the small displacement domain leading the identified contact points of a shape on the contact points of the other one.

This paper is composed of three main parts. The first one presents the method for the relative positionning of parts regarding their form devition and a contact force direction. The second part presents the random generation of form deviation using the modal paramterization and the translation of the form deviation in two geometrical parameters. The third part presents the case of application that is a 2D linear linkage. A single study is firstly presented, and then simulations are drawn with differents values of form deviations. The paper ends by a conlusion.

2 Proposed Method

The assembly of one part on another is illustrated by the figures 1 and 2. Figure 1a shows two shapes with form deviation facing each other. Shapes are theoretically positioned and ideal shapes are associated by the least square criterion. Figure 1b shows the first step of the method that transfers the upper shape form deviation to the lower one. The transferred deviation, also called distance surface, represents the point to point distances between the shapes. Shapes are in contact if two distances in a 2D problem are set to zero. Possible contact points and facets are identified thanks to the computation of the convex surface that filters the possible contact points. Figure 1b shows two different force directions, force a and force b, used to identify the contact facet and its contact points.

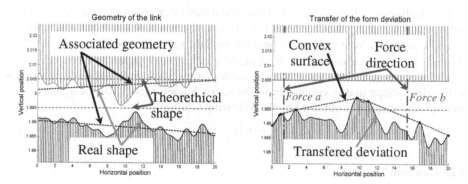

Fig. 1. a) two shapes with form deviation, b) transferring one shape form deviation on the other shape

The following figure 2 shows two different positioning, figure 2a and 2b, depending on the force direction of force a and force b respectively.

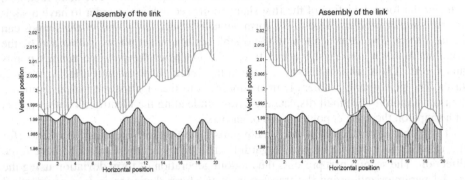

Fig. 2. Two contact configurations of shapes positioning, a) assembly for the direction of the given force a, b) assembly for the direction of the given force b

3 Parameterization of Shapes

This part presents two ways to parameter the form deviation. The first way is a description of the form deviation by elementary form deviation with a proposed method called the modal parameterization. The second way uses much simpler parameters that are geometrical zones.

3.1 Modal Parameterization

Samper [13-14] uses the modal analysis to generate a form deviation basis to characterize a measured form deviation. The main advantage of this method is that any type of geometry and form deviation can be characterized as a combination of elementary form deviation. The following figure 3 shows the height first modes of a 2D linear shape.

The analysis of a form deviation in the modal basis is a vectorial projection in a non orthogonal basis using the dual basis [15]. Hence, the result of the modal characterization of a measured form deviation V on the B modal basis is the modal signature Λ. This Λ modal signature is composed of the λ_i modal coefficients with metric meaning.

The recomposed shape with the rigid modes (translation and rotation) corresponds to the rigid shape and is equivalent to the Least Square associated ideal shape. The recomposed shape R is obtained by the following relation:

$$R = \Lambda.B \qquad (1)$$

3.2 Random Shapes

If a batch of produced part exists, then it is possible to draw simulations based on this family of form deviations. In other cases, assumptions can be made in order to define

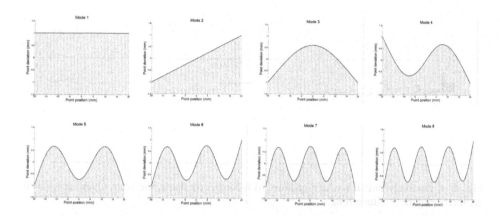

Fig. 3. First height modal vectors of a 2D profiles

the family shapes (batch of form deviations). The modal characterization of form errors can be used to create simulated shapes. A random draw of the modal coefficients creates a random shape which form complexity depends on the number of considered modes. Based on the observation of shapes analyses, amplitude of the modal coefficients is considered given by the following law:

$$A(i) = A_0 / i \qquad (2)$$

Where A_0 is initial amplitude, i is the order of the modal coefficient and $A(i)$ is the maximum amplitude of the ith modal coefficient. The following figure 4a shows the amplitude law of the coefficients and a random draw of a modal signature. This decreasing law can be changed in order to fit to a pilot production.

3.3 Geometric Parameters

The modal parameterization presented in this paper is only used to generate random shapes (lines in this case). Then two parameters are calculated, closer to geometrical specification:

- The strength of the shape that indicates the range of deviation,
- The localisation deviation of an ideal shape associated by the least square method.

The figure 4 represents the random drawing of the modal signature and the corresponding shape. On this shape is calculated the strength and the deviation of the associated geometry.

For this shape, the rectitude zone is 10.5µm, the localisation deviation of the associated geometry is 3.8µm and the localisation deviation of the entire shape (including the form deviation) is 8.0µm.

In order to identify the influence of parameters, shapes are randomly generated with modal signatures, then strength and localisations are measured. It is then possible to modify the modal signature thanks to an algorithm in order to set a given value to the shape rectitude and/or to the localisation of the associated geometry.

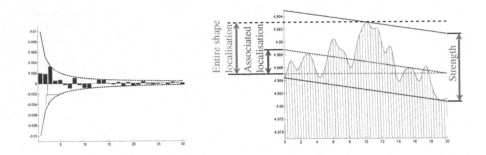

Fig. 4. a) randomly drawn modal signature, b) recomposed shape, the least square associated geometry and the rectitude zone

4 Clearance Domain with Form Deviation

This part presents our proposed method to find the clearance domain of a mechanical linkage regarding their form deviations based on the method detailed in the first part. A study case illustrates the method on a single application, and then simulations are drawn to compare our approach to the least square assembling model.

4.1 2D Linear Linkage

To illustrate this paper, the study case is a 2D linear linkage, composed of two 2D linear contacts, upper and lower contacts. For this example, the geometrical clearance of the linkage is set to 20μm and form deviation are randomly drawn then measured.

Based on our proposed methods, possible contact facets and points are identified for each contact area, on the upper and lower.

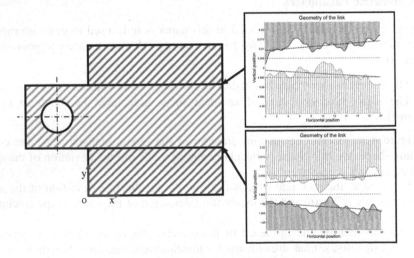

Fig. 5. Tthe mechanical linkage and its upper and lower contact zones

Table 1. Values characterizing the shapes of the linkage

	Strength (μm)	Localisation* (μm)	Real localisation (μm)
Shape 1	11.0	4.8	11.8
Shape 2	10.5	3.8	8.0
Shape 3	12.4	3.5	9.1
Shape 4	9.3	4.2	11.2

* *deviation of the localisation of the least square associated geometry*

An added difficulty is the fact that the linkage can not be assembled due interpenetration of parts. Hence, for each relative positioning of the inner part on the outer one, interpenetration is checked. The following figure 6 illustrates a positioning with interpenetration of parts; this relative positioning is no considered.

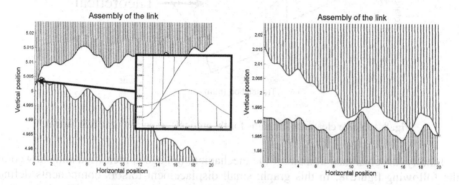

Fig. 6. Non-possible position of the linkage due to interpenetration on the upper contact, a) upper contact zone with interpenetration, b) lower contact zone with contact

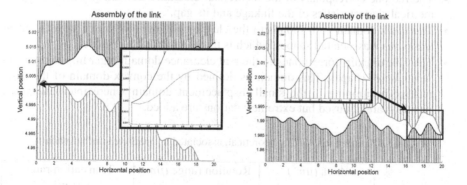

Fig. 7. One extreme rotation in the linkage, a) upper contact zone with contact, b) lower contact zone with contact

Another difficulty is to find the two relative positioning given by the extreme rotations. An iterative approach is chosen to find both extreme rotations in the linkage. The following figure 7 shows one extreme rotation where contact points are not located at the extremity of the shapes.

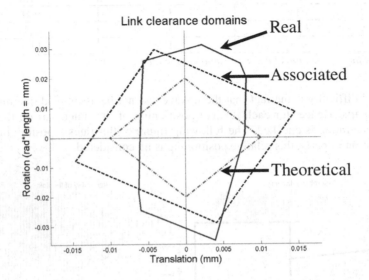

Fig. 8. Linkage clearance domains for theoretical, associated and real model

Finally, the clearance domain of the mechanical linkage is identified and showed on the following figure 8. In this graph, small displacement torsors components define the axis. Limits are always convex (small displacement assumption). Three domains are obtained:

- The red one corresponds to the theoretical clearance domain given by the geometrical characteristics of the linkage and its gap,
- The black domain corresponds to the clearance domain obtained with the ideal least square associated shapes, which is the commonly used model,
- The blue domain corresponds to the real clearance domain regarding to the form deviation of shapes. This domain is defined by the convex domain of the parts relative positioning in the small displacement domain where interpenetration positions are not used but extreme rotations are added.

Table 2. Characteristics of the theoretical, associated and real clearance domains

	Area (μm^2)	Rotation range (μm)	Translation range (μm)
Theoretical	400	40	20
Associated	925 (= 231% *)	58.6 (= 147% *)	29.3 (= 147% *)
Real	755 (= 189% *)	66 (= 165% *)	14.5 (= 73% *)

* compared to the theoretical clearance domain

To compare these different clearance domains, it is chosen to measure the domains area, translation range (horizontal length) and rotation range (vertical length). The following table 2 shows the different domains measures where it can be observed that, for this particular case, the theoretical domain area is more or less half of the associated and real domains areas. Another remark is that the associated domain is almost homothetic compared to the theoretical one, but the real domain shows that although the rotation range is a bit larger than the translation range of the associated domain, the translation range is half of the translation range of the associated domain.

4.2 Simulations

This part presents simulations in order to evaluate the influence of the strengthness and associated localization zones. The clearance value of the mechanical link is set to 6μm. Strengthness and associated localization of shapes are independantly set from 0 to 12μm by step of 2μm. For each configuration, 1000 assemblies are drawn, and their associated and real clearance domains are measured. Hence, the following results are made by 7*7*1000 simulations.

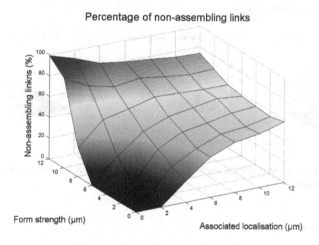

Fig. 9. Evolution of the rate of non-assembling linkages depending on the forms strengths and associated localisation for the 6μm linkage clearance value

The first result showed in the figure 9 is the rate of non-assembling linkage given the deviation zones of shapes find with the real model considering the form deviation. It can be observed that almost all linkage can be assembled when the associated zone is null and the shape strength is lower or equal to the linkage clearance.

The next figures concern the dimensions (area, translation and rotation range) of the associated and real clearance domains. Plotted surfaces represent the mean dimensions of the clearance domains of the drawn assemlies.

It can be observed on the following figure 10 that the domains dimensions are identical to the theoretical domains when no associated deviation are considered, this is due to the model that does not consider form deviation. Another remark is that dimensions of the domains grow as deviation zones grow, there is no compensation due to the strengthness it can be observed on our model as showed in the following figure 11.

The figure 11 c shows that when the associated deviation zone equals or is upper to the linkage clearance, the strength almost has no effect on the rotation range of the clearance domain. However, as the strength increases the real clearance domains characteristics deacrease.

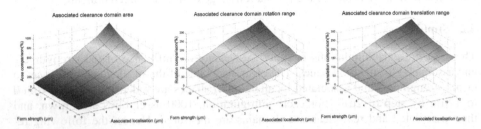

Fig. 10. Evolutions of the associated clearance domains, a) domain area, b) domain rotation range, c) domain translation range

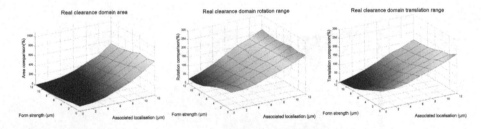

Fig. 11. Evolutions of the real clearance domains, a) domain area, b) domain rotation range, c) domain translation range

5   Conclusion

This paper continues our previous works dealing with form deviations in assemblies and shows that form deviations of parts not only impact their relative positioning but also impact the assembly clearance domain of the mechanical linkage.

It can be conclude that the rigid assembly model usually considered for tolerancing is not correct concerning the assembly prediction with form deviation, even in a small deviation zone. Hence the solution proposed by [6] and others, the strength deviation is considered as an additional rigid deviation zone, can appear to be a correction of this problem.

Then, works have to be continued to evaluate more precisely the separated and combined influence of each form deviation on the assembly functional requirement. The aim will be to find an expression or a criterion for the acceptance of the shape deviation that guarantees the functional requirement.

References

1. Graves, S.: Tolerance Analysis Formula Tailored to Your Organization. Journal of Quality Technology 33(3) (2001)
2. Chase, K.W.: Tolerance Analysis of a 2D and 3D assemblies, ADCATS Report, No 94-4 (1999)
3. Davidson, J.K., Shah, J.J.: Geometric Tolerances: A New Application for Line Geometry and Screws. ImechE Journal of Mechanical Engineering Science 216, Part C, 95–104 (2002)
4. Bourdet, P., Mathieu, L., Lartigue, C., Ballu, A.: The Concept of Small Displacement Torsor in Metrology. In: Proceedings of the International Euroconference, Advanced Mathematical Tools in Metrology, Oxford (1995)
5. Giordano, M., Kataya, B., Pairel, E.: Tolerance Analysis and Synthesis by Means of Clearance and Deviation Spaces. In: Proceedings of the 7th CIRP International Seminar on Computer Aided Tolerancing, pp. 345–354 (2001)
6. Ameta, G., Davidson, J.K., Shah, J.: Influence of Form Frequency Distribution for a 1D Clearance Wich is Generated from Tolerance-Maps. In: Proceedings of the 10th CIRP Conference on Computer Aided Tolerancing, Specification and Verification for Assemblies, Erlangen Germany (2007)
7. Radouani, M., Anselmetti, B.: Contribution à la Validation du Modèle des Chaînes de Cotes – Etudes Expérimentale du Comportement de la Liaison Plan sur Plan. In: Congrès International Conception et Production Intégrées, Meknès Moroco, October 22-24 (2003)
8. Neville, K.S.L., Grace, Y.: The modeling and analysis of butting assembly in the presence of workpiece surface roughness part dimensional error. International Journal of Advanced Manufacturing Technologies 31, 528–538 (2006)
9. Stoll, T., Wittman, S., Meerkamm, H.: Tolerance Analysis with detailed Part Modeling including Shape Deviations. In: Proceedings of the 11th CIRP Conference on Computer Aided Tolerancing, Geometric Variations within Product Life-Cycle Management, Annecy France (2009)
10. Morière, S., Mailhé, J., Linares, J.-M., Sprauel, J.-M.: Assembly method comparison including form defect. In: Proceedings of the 11th CIRP Conference on Computer Aided Tolerancing, Geometric Variations within Product Life-Cycle Management, Annecy France (2009)
11. Adragna, P.-A., Samper, S., Favrelière, H., Pillet, M.: Analyse d'un assemblage avec prise en compte des défauts de forme. In: Congrès International Conception et Production Intégrées, Rabat Morocco (2007)
12. Adragna, P.-A., Favrelière, H., Samper, S., Pillet, M.: Statistical assemblies with form errors – a 2D example. In: Proceedings of the 4th International Precision Assembly Seminar (IPAS 2008), Micro-Assemblies Technologies and Applications, Chamonix France, pp. 23–33 (2008)
13. Formosa, F., Samper, S., Perpoli, I.: Modal expression of form defects. In: Models for Computer Aided Tolerancing in Design and Manufacturing, pp. 13–22. Springer, Heidelberg (2007)
14. Adragna, P.-A., Samper, S., Pillet, M.: Analysis of Shape Deviations of Measured Geometries with a Modal Basis. Journal of Machine Engineering: Manufacturing Accuracy Increasing Problems – Optimization 6(1), 95–102 (2006)
15. Favrelière, H., Samper, S., Adragna, P.-A., Giordano, M.: 3D statistical analysis and representation of form error by a modal approach. In: Proceedings of the 10th CIRP Conference on Computer Aided Tolerancing, Specification and Verification for Assemblies, Erlangen Germany (2007)

PART II

Micro-assembly Processes

Chapter 3

Development of Micro-joining Processes

Precision Assembling and Hybrid Bonding for Micro Fluidic Systems

Agathe Koller-Hodac, Manuel Altmeyer, and Silvio Walpen

Institute for Laboratory Technology, University of Applied Sciences Rapperswil,
Oberseestrasse 10,
CH-8640 Rapperswil, Switzerland
{agathe.koller,manuel.altmeyer,silvio.walpen}@hsr.ch

Abstract. An automated process is described to assemble highly functionalized micro fluidic cartridges. Different materials such as glass, silicon and thermoplastics can be bonded using adhesive gaskets. The assembling and bonding sequence is realized at room temperature so that the biological function of the cartridge is fully preserved. Several control tests for alignment and orientation are integrated into the process so that a precise and reliable assembly using robotic equipment can be achieved.

Keywords: Micro-assembly, hybrid bonding, micro fluidics, micro-gripper.

1 Introduction

Miniaturization of biological assays is a major trend in the pharmaceutical and life science industry. This allows not only the reduction of expensive reagents, but also enables the integration of additional functions such as mixing or dosing into micro fluidic cartridges. Unfortunately the integration of such advanced functionalities fails in production due to the lack of suited reliable automated assembly technologies that are also able to maintain the biological function of the micro fluidic device.

Due to increasing demand for optical and micro-electro-mechanical Systems, micro-assembly platforms and tools have been developed in recent years [1]. Considering the bonding process in an industrial environment, interesting approaches have been reported in the semiconductor sector. The automated bonding is mainly based on a thermal process.

In life sciences, especially in the micro fluidic domain, conventional automated processes are not applicable to the bonding of micro fluidic devices when biomolecules must be deposited onto the chip prior to sealing. Also new assembly issues must be addressed that are often not compatible with existing processes. Until now, assembly of complex micro fluidic components made of different materials is done either manually or semi-automatically by using a range of heterogeneous tools such as microscopes, precision stages or fixtures. However, this approach suffers from a lack of flexibility, scalability and reproducible quality. The absence of appropriate equipment prevents a mass production and a cost-effective commercialization of micro fluidic components. Therefore, there is an increasing need for innovative and low-cost processes for automated high precision assembly of hybrid micro fluidic components, such as plastic and glass.

S. Ratchev (Ed.): IPAS 2010, IFIP AICT 315, pp. 65–72, 2010.

Our main research and development focus lies on a fabrication process which maintains the biological function throughout the entire automated assembly of the micro fluidic product. The highly functional cartridges require the integration of additional specialized elements into the plastic micro fluidic device, such as chemically functionalized detection units made of silicon or glass. Reliable hybrid bonding is becoming the key issue in cartridge production. Biocompatibility, low temperatures and no out-gassing of any material in the cartridge are just few points that have to be fulfilled. In addition the assembly and bonding process must be cost efficient and adapted to different cartridge geometries.

This paper presents an automated assembly process for multi-layer micro fluidic cartridges consisting of a scaffold, a double-sided adhesive gasket and a pre-coated chip. Special emphasis has been on fully preserving function of pre-coatings during the bonding process. No heat or ultraviolet light is necessary during assembly. For a systematic evaluation of the process, alignment monitoring as well as quality control of the final device have been developed.

2 Assembly Process and Layout

In the first section, the different layers of micro fluidic cartridge are introduced. In the second section, the sequential steps of the assembly process are highlighted. Special emphasis lies on the correct orientation and precise alignment of components before assembling. Finally, the quality control of the final cartridge is described.

2.1 Cartridge Components

Components of the micro fluidic cartridge have been designed to facilitate the assembly process. Several alignment features are implemented on the scaffold and the adhesive gasket and its protective foil have been specially developed to support the automated assembly. Type and thickness of the gasket material turned out to be essential parameters to achieve high bonding quality. The glass chip has a quadratic shape with no reference marks and is pre-coated with biomolecules.

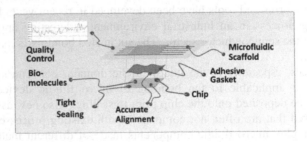

Fig. 1. Components of micro fluidic cartridge

Scaffolds and chips are supplied on trays which are placed within the robotic cell. Supply of adhesive gaskets requires special equipment and methods, since double-sided tape is very delicate to handle. The developed solution is based on an automated feeding system with foil rolls.

2.2 Precision Assembling

A major challenge associated to this development is to perform reliable and accurate assembling of micro fluidic parts without requiring an expensive apparatus. The production cell, as shown in Fig. 2, includes several mounting modules as well as storage trays for scaffolds and chips. The size and complexity of the assembling and hybrid bonding processes made it necessary to define sub processes and integrate quality control checks.

Fig. 2. The robotic cell includes an articulated robot (middle), two supply systems for chips and scaffolds (left and right of the robot), a feeding system for adhesive gaskets (front) and a sealing test module (front left)

The experimental setup includes a standard articulated industrial robot with six degrees of freedom for coarse positioning tasks. The robot kinematics with six rotational axes allow placing chips and especially scaffolds in a number of positions and orientations required to perform assembling and testing operations. Moreover, the articulated robot Kuka KR3 has an adequate workspace to supply and storage sufficient parts for long production autonomy. Fine positioning is achieved using a hybrid gripper, which design allows normal pick and place operations as well as accurate joining of several micro fluidic part layers. A vacuum system allows grasping chips at the outside edge, while entirely preserving surface coating with bio-molecules.

The Fig. 3 describes the complete assembling process which has been split into sub processes. All cartridge components are supplied and automatically mounted in the robotic cell. Exact position or orientation of each component is controlled before each assembling step.

The robot picks the parts out of the trays and performs initial handling tasks. Since transparent chips have an asymmetric coating, their orientation has to be detected for correct assembling. A standalone vision system identifies the chip orientation and transmits data to the robot control. A dedicated module for automatically supplying and preparing adhesive gaskets has been developed and integrated into the robotic cell.

The assembling quality of the micro fluidic cartridge plays an important role in the application assay. In particular, a part alignment accuracy better that 20 μm is

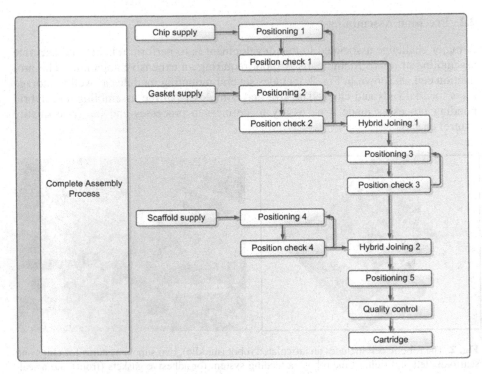

Fig. 3. Description of the assembly process: Exact component position is controlled before each hybrid bonding step

required, which is higher than the accuracy of ±0.05 mm provided by the robot [2]. Therefore, the first joining step requires position monitoring based on a control loop consisting of vision sensor, image processing software and robot control. This control algorithm is based on the look-and-move principle [3], [5]. After this first joining step, chip and gasket are precisely positioned on the hybrid gripper.

The second joining step combining chip and gasket with the scaffold is solved using the mechanical mounting support of the hybrid gripper. Its guiding balls interact with the scaffold grooves, leading chip and gasket to their correct position on the scaffold. In the last process step, the assembling quality of the micro fluidic cartridge is assessed using a sealing check module. The critical steps are described in details in the following sections.

2.2.1 Chip Orientation Detection

The glass chip consists of a quadratic plate without additional mechanical or optical orientation marks. This must be correctly orientated due to pre-coating asymmetry. Placed in the inspection cradle as shown on Fig. 4, the chip has eight possible orientations: Four ninety degree turning steps and front- or backside. The chip being the most valuable part of the assembly, a full orientation control is required to ensure a correctly assembled cartridge at a minimal yield.

Fig. 4. Chip Inspection Cradle

The coating on the crystal clear glass substrate is highly translucent. The coating pattern reflects green whereas the background appears violet. A reliable and stable detection has been obtained by using a color camera system.

Based on the preprocessed camera information, the image processing algorithms can detect the ninety degree rotation and can differentiate between front and back sides. Following the concept of decentralized modules, a smart camera which contains both camera and image processing is used here.

2.2.2 Gasket Supply
A major step in the assembly process is the supply and handling of double-sided adhesive gaskets. Besides the tape material and thickness, the supply procedure is essential to obtain reliable bonding results. Based on preliminary tests, the gasket supply on an endless carrier foil covered by an additional protection foil has proven to be the best solution. Perforations like on 35 mm cinema films or analog picture films have been laser-cut at the side of carrier foils to provide reference marks.

The tape feeding system is shown on Fig. 5. The bonding area for gasket and chip is located on a glass plate (A). The roller of small diameter (B) has notched dents matching the perforations of the carrier foil. Being driven by a controlled stepper

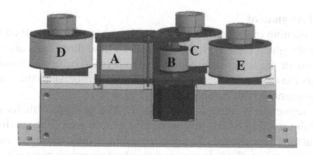

Fig. 5. The tape feeding system supplies adhesive gaskets automatically. Gaskets are provided on foil rolls and are covered by protection foils. Before hybrid bonding, one gasket is moved to the joining area while the protection foil is removed.

motor, it is responsible for the gasket basis positioning. The spender roller (C) in the center unwinds the adhesive gasket. The bobbin (D) reels the blank carrier foil after chip and gasket bonding is performed in the joining area. The bobbin (E) reels the protection foil. All these three rollers are motor driven to ensure the correct foil tension at the bonding position and to avoid uncontrolled unwinding.

2.2.3 Chip Alignment

The chip has been grasped by the robot vacuum gripper and the adhesive gasket has been provided by the tape feeding system. The next process step is the delicate assembling of the chip to the adhesive gasket. A vision system detects the exact position of the adhesive gasket through the glass plate of the tape feeder. Based on a calibrated picture, the position offset of the gasket is determined and provided to the cartesian controller of the robot. The control algorithm illustrated in Fig. 6, based on the look-and-move principle, checks and eventually corrects the position of gripper relative to the adhesive gasket, before the chip is bonded to the gasket.

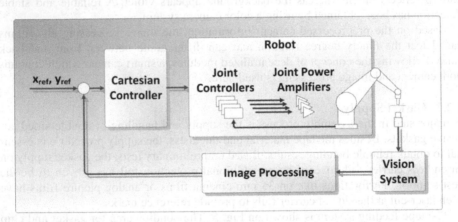

Fig. 6. Look-and-move control loop: a vision system provides exact information about gasket position. The robot aligns the chip to the adhesive gasket before hybrid bonding.

2.2.4 Scaffold Alignment

The precise positioning of the scaffold with respect to the chip is solved mechanically using guiding balls glued onto the hybrid gripper. These three balls are aligned to the grooves designed in the injection molded scaffold [4]. The combination of grooves and balls hinders the degrees of freedom of the scaffold relative to the gripper, leading to a precise alignment of both components.

The joining sequence is illustrated in Fig. 7. The gripper with its guiding balls moves close to the scaffold which lies on an air cushion area. The scaffold is floating when the air cushion is activated. The groove arrangement results in an exact alignment between chip and scaffold. In a next step, the gripper element which holds chip and gasket moves linearly downward the scaffold and joins both components so that the micro fluidic cartridge is finally assembled.

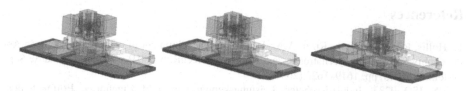

Fig. 7. Assembling sequence for chip and scaffold: (1) the robot gripper gets close to the scaffold, (2) the scaffold is floating on an air cushion (3) the gripper is actuated linearly in order to bond chip to scaffold

2.3 Quality Control

Micro fluidic cartridges have to be mounted with high precision due to small dimensions of micro fluidic channels and interfaces. As already described, several modules have been integrated into the robotic cell to control and adjust position, alignment and orientation during the assembly process. An additional module performs automatically a sealing test of the final cartridge. This test consists in checking pressure in all micro fluidic channels. Assembled cartridges which fail the sealing test have been discarded.

3   Conclusion

An innovative yet reliable method has been achieved for automatically assembling micro fluidic parts composed of several layers.

A robotic cell has been developed and realized to assemble highly functionalized micro fluidic cartridges in an industrial environment. Experimental results have shown that hybrid bonding of different materials - such as micro-structured plastic parts and high-quality, bio-functionalized silicon or glass elements - can be performed using adhesive gaskets. The whole assembling and bonding process is realized at room temperature so that the biological function of the cartridge is preserved.

While the whole automated process is based on a standard industrial robot, the precise assembly of multiple layers has been made possible through an innovative gripper design. Further, a vision based look-and-move control approach allows the robot to adjust position relative to micro fluidic layers, in order to precisely align micro fluidic channels to pre-coated spots.

Acknowledgement

Our sincere thank to Markus Lüthy, Edi Krüttli und Kurt Eggmann from Weidmann Plastics Technology AG for their support and part manufacturing.

Grateful thanks also to Max Wiki from Dynetix AG and to Stephane Follonier, Janko Auerswald und Qun Lai from CSEM SA for their expertise and important contribution to the project.

This project has been supported by the Swiss Innovation Promotion Agency (CTI).

References

1. Hollis, R.L., Gowdy, J., Rizzi, A.A.: Design and Development of a Tabletop Precision Assembly System. In: Mechatronics and Robotics (MechRob 2004), Aachen, Germany, September 13-15, pp. 1619–1623 (2004)
2. EN ISO 9283, Industrieroboter, Leistungskenngrössen und zugehörige Prüfmethoden. Beuth-Verlag, Berlin (1999)
3. Conticelli, F., Allotta, B.: Two-Level Visual Control of Dynamic Look-and-Move Systems. In: IEEE International Conference on Robotics & Automation, San Francisco, CA (2000)
4. Wiendahl, B., Lotter, H.P.: Montage in der industriellen Produktion - Ein Handbuch für die Praxis. Springer, Heidelberg (2006)
5. Schöttler, K., Raatz, A., Hesselbach, J., Wu, H.: Size-adapted Parallel and Hybrid Parallel Robots for Sensor Guided Micro-Assembly. In: Parallel Manipulators; Towards New Applications, pp. 225–244. I-Tech Publication and Publishing (2008) ISBN 978-3-902613-40-0
6. Schöttler, K., Raatz, A., Hesselbach, J.: Precision Assembly of Active Microsystems with a Size-Adapted Assembly System. In: Fourth International Precision Assembly Seminar (IPAS 2008), Chamonix, France, pp. 199–206. Springer, Boston (2008)

Feasibility of Laser Induced Plasma Micro-machining (LIP-MM)

Kumar Pallav and Kornel F. Ehmann

Department of Mechanical Engineering
Northwestern University
Evanston, IL 60208, USA
kumarpallav2008@u.northwestern.edu, k-ehmann@northwestern.edu

Abstract. The paper offers evidence on the feasibility of a newly proposed micro-machining process, which is motivated by the need to overcome the various limitations associated with μ-EDM and conventional ultra-short laser micro-machining processes. The limitations in μ-EDM and laser micro-machining processes are mainly due to the requirement of a conductive electrode and workpiece, electrode wear and compensation strategies, and complex process control mechanisms respectively. The new process uses a laser beam to generate plasma in a dielectric near the workpiece surface whose explosive expansion results in material removal by mechanisms similar to those that occur in μ-EDM.

Keywords: Laser-machining, Micro-machining, μ-EDM, Plasma.

1 Introduction

Micro-manufacturing encompasses the creation of high precision 3D products with feature sizes ranging from a few microns to a few millimeters. While micro-scale technologies are well established in the semiconductor and micro-electronics fields, the same cannot be said for manufacturing products involving highly accurate complex 3D geometries in materials such as metals, ceramics and polymers.

Micro-EDM (μ-EDM) has been one of the frontrunners for material removal in the aforementioned size range. Material removal is facilitated by the plasma created when the electric field generated due to high voltage between the electrode and the workpiece exceeds the threshold value for dielectric breakdown. The created plasma explosively expands and sends shock waves to the surface of the workpiece at very high speed that lead to localized heating and ablation. It has been widely recognized that the major limitations of μ-EDM are [1-3]: (1) the inability to micro-machine non-conductive materials and metals with low conductivity, (2) non-productive time and cost associated with on-machine electrode manufacturing, (3) frequent electrode replacements due to high electrode wear, (4) flushing of debris from the small electrode-workpiece gap (< 5 μm), and (5) low MRR (0.6 - 6 mm^3/hr).

A competing process is ultra-short laser pulse-based micro-machining. Ultra-short pulsed lasers have pulse durations ranging from (ps) to few (fs) at pulse frequencies ranging up to 500 kHz, and very low average power (up to 10 W). However, since

S. Ratchev (Ed.): IPAS 2010, IFIP AICT 315, pp. 73–80, 2010.

pulse duration is very short, the peak power is very high [4, 5]. With ultra-short laser pulses, the ablation process is dependent on a wide number of parameters including the type of polarization, type of material, peak pulse power, pulse repetition rate, pulse duration, environmental conditions and pressure, and wavelength. The ablation process is also accompanied by electron heat conduction and formation of a molten zone inside the metal target which reduces precision when machining metals [7]. The dependence of the ablation process on so many parameters makes conventional ultra-short laser micro-machining a very complex process and controlling all of these parameters is a major challenge [7].

The new micro-machining process described in this paper is motivated by the need to overcome the limitations of μ-EDM and of conventional ultra-short laser micro-machining. The new micro-machining process, to be termed *Laser Induced Plasma Micro-machining* (LIP-MM) uses an alternative mechanism for plasma creation to facilitate material removal.

1.1 The Laser-Induced Plasma Micromachining (LIP-MM) Process

The underlying principle of the proposed LIP-MM process rests on the fact that when an ultra-short pulsed laser beam is tightly focused in a transparent dielectric medium, extremely high peak power densities are reached at the focal spot [6, 8, 9]. And, when the peak power density exceeds the threshold irradiance required for dielectric break-down, rapid ionization of the medium occurs leading to plasma formation. The plasma formed is optically opaque and causes a drastic increase in the absorption coefficient, which, in turn, gives rise to a rapid energy transfer from the radiation field to the medium [6, 8]. This process is, in essence, an optical breakdown when the free electron density at the focal spot exceeds the critical value of 10^{20} cm^{-3}. At this value, the plasma absorbs nearly all of the incoming laser beam's energy leading to rapid heating of the material in the focal volume [9]. The plasma leads to a liquid-gas phase change and causes cavitation accompanied by the generation of shock waves. If the focal volume is slightly above the surface of the workpiece, the explosive expansion of the plasma that is followed by the generation of shock waves, like in the case of a μ-EDM discharge, results in material removal.

Plasma generation by ultra-short pulsed lasers, as described above does not require a conductive electrode and workpiece [9]. Actually, there is no need for an electrode. Even more importantly, the plasma's properties can be controlled by controlling the laser process parameters. Hence, many of the limitations of μ-EDM and conventional laser micro-machining can be potentially circumvented through the LIP-MM process. In the subsequent sections initial results aimed at confirming the viability of the proposed material removal mechanism are presented.

2   Experimental Setup and Results

For the experimental realization of the LIP-MM process, a commercial, Nd:YVO$_4$ laser system (Lumera laser, Rapid) emitting laser pulses of 8 ps pulse duration at a 532 nm wavelength was used. The beam profile of the (ps) pulses was Gaussian (TEM$_{00}$) and the beam was linearly polarized. The pulse repetition rate for the laser system varied from 10 kHz to 500 kHz. For plasma generation, since very high peak power densities are required, low repetition rates ranging from 10 kHz to 50 kHz were used. The laser machine used in the experiments is capable of producing a peak power density level of several 10^{11} W/cm^2 at a spot size of 21 µm in diameter. It is also equipped with a 5-axis programmable motion controller that directs the beam. The system has a resolution of 0.01 µm and feed rates ranging from 0.01 µm/s to 10 cm/s, allowing a very precise control of the relative motion between the beam/plasma and the workpiece.

The laser beam was brought to the focusing lens through a beam delivery system consisting of various mirrors and a 3X beam expander. The beam expander expands the beam diameter from 0.6 mm to 1.8 mm before it reaches the focusing triplet lens. The focusing triplet lens focuses the laser beam to a diffraction limited spot size of 21 µm at the focal spot. Water was selected as the dielectric for the LIP-MM process. The laser pulses were focused into a glass beaker filled with water. Figure 1 shows the experimental setup used in the LIP-MM process.

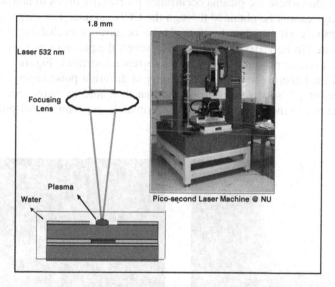

Fig. 1. Schematics of the LIP-MM Process

2.1 Determination of Breakdown Threshold and Plasma Occurrence Probability

The peak power density reached at the focal spot with ultra-short lasers generally ranges from several 10^9 W/cm^2 to few 10^{11} W/cm^2. Ionization starts as soon as the peak power density exceeds the threshold ionization potential for the dielectric medium that depends on a number of factors including: laser wavelength, laser and dielectric media interaction time, and additional conditions such as the number of seed electrons, the type of dielectric, and the type of polarization of the laser beam [9]. Dielectric breakdown resulting in plasma generation can only be achieved when the peak power density exceeds the threshold value required to initiate ionization [6].

For plasma generation and control, the breakdown threshold corresponding to different pulse repetition rates was determined by imaging the plasma formed with a CCD camera (Edmund Optics, Guppy F-146 FireWire A 1/2" CCD Monochrome Camera) with a 2.5 - 10X magnification (Edmund Optics, VZM 1000i Video Lens) as the peak power density was varied from superthreshold to subthreshold values. During each laser exposure, the output average power of the incident laser irradiation was measured using an external laser power meter (Gentec SOLO 2 (R2)). Plasma generation due to dielectric breakdown was visually detected in a darkened room with the CCD camera and was recorded at a frame rate of 15 frames per second. Figure 2 shows the plasma images obtained through the CCD camera at different average power levels and pulse repetition rates. For each peak power density value at every pulse repetition rate, at least 20 laser exposures were evaluated for determining the plasma occurrence probability.

As shown in Fig. 2, with a decrease in peak power density levels at all repetition rates, the intensity of the plasma image decreases until it reaches the dielectric ionization threshold value where the plasma occurrence probability drops to below 10% and the plasma image cannot be obtained through the CCD camera. The mean breakdown threshold intensity corresponding to 10% plasma occurrence probability was experimentally measured to be 1.10×10^{11} W/cm^2. However, the plasma size remained independent of the average power levels at all pulse repetition rates. Figure 3 shows the dependence of the plasma occurrence probability at different pulse repetition rates on peak power density. It shows that the breakdown threshold in water, corresponding to 10% plasma occurrence probability is approximately constant for all pulse repetition rates.

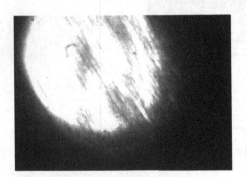

Fig. 2. Plasma Images at 10 kHz, 63 mW and 38 mW Average Power Levels

2.2 Laser Induced Plasma Micro-machining of Steel and Quartz

For making micro channels in steel using the LIP-MM process, a pulse repetition rate of 10 kHz and a peak power density corresponding to a 70% plasma occurrence probability were selected. The laser beam traverse speed was kept at two levels, i.e., at 100 μm/sec and 500 μm/sec for machining single pass channels.

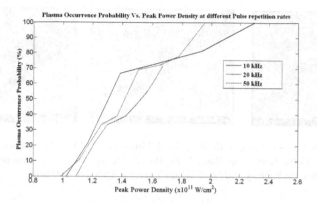

Fig. 3. Plasma Occurrence Probability at Different Peak Power Densities and Pulse Repetition Rates

Figure 4 shows the image of a single pass channel made in steel with the LIP-MM process at a laser beam traverse speed of 100 μm/sec and peak power density corresponding to 70% plasma occurrence probability at a 10 kHz pulse repetition rate. The channel has an approximately 30 μm depth at various cross-sections and the walls of the channel are not vertical. The depths of the micro-machined channels were examined at various locations through a white light interferometer (Zygo NewView 7300). It was observed that the depth of the channels varied inversely with variations in the laser beam traverse speed. At a lower laser beam traverse speed of 100 μm/s, the depth of the channel was approximately 30 μm, whereas with the higher laser beam traverse speed of 500 μm/s, the depth of the channel was only 6 μm.

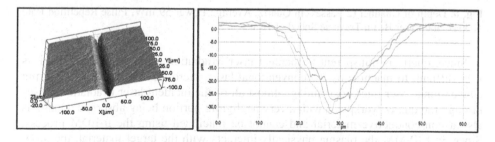

Fig. 4. Channels Made by LIP-MM in Steel at an Average Power of 59 mW, Pulse Repetition Rate of 10 kHz and Laser Beam Traverse Speed of 100 μm/s

Figure 5 shows the images obtained through a SEM (Hitachi S3400N II) of channels machined in steel with different numbers of passes. The other process parameters were similar to the channel shown in Fig. 4.

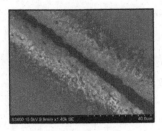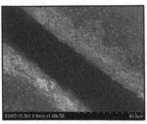

Fig. 5. SEM Images of Channels Made in Steel Through LIP-MM Process with 1, 5 and 10 Passes, 10 kHz Pulse Repetition Rate, Peak Power Density Level Corresponding to 70% Plasma Occurrence Probability and Laser Beam Traverse Speed of 100 μm/s

As shown in Fig. 5, with single pass micro-machining the channel had a significant recast layer and inclined walls. However with an increase in the number of passes, the walls became vertical and the recast layer decreased significantly. The depth of the channel became more uniform and also increased with an increase in the number of passes. Hence, with an increase in the number of passes, the surface features of the channel improve due to the vertical walls, uniform depth, and significant reduction in the recast layer.

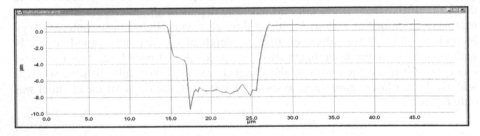

Fig. 6. Depth of Channel (2 Passes) in Quartz, Average Power 59 mW, Pulse Repetition Rate 10 kHz, and Laser Beam Traverse Speed of 1μm/s

Since, quartz is transparent to incident laser irradiation at most wavelengths, it is very difficult to machine by the conventional laser micro-machining process. During conventional laser micro-machining, almost all of the incident laser irradiation passes through the quartz workpiece with a very negligible portion being absorbed. Quartz is also a non-conductive material and cannot be machined using the μ-EDM process. Since, in LIP-MM the plasma physically interacts with the target material, the problem of negligible absorption is not encountered.

Single pass and multiple pass channels were made in a quartz workpiece by the LIP-MM process at different peak power density levels and laser beam traverse speeds. Figure 6 shows the depth of a typical multiple pass channel made in quartz, measured through an interferometer (Zygo NewView 7300). The dependence of the depth of a single pass channel on laser beam traverse speed at different peak power density values in quartz is depicted in Fig. 7. It can be seen that irrespective of the peak power density level, significant reduction in channel depth occurs with an increase in the laser beam traverse speed.

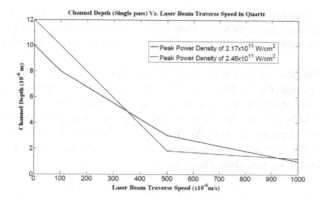

Fig. 7. Dependence of Channel Depth (Single Pass) in Quartz on Laser Beam Traverse Speed at Different Peak Power Density Levels

3 Results and Discussion

The breakdown threshold for water corresponding to 10% plasma occurrence probability was found to be equal to 1.10×10^{11} W/cm^2. The peak power density depends on the average power, pulse repetition rate, pulse duration, and the focal spot size. With 8 ps pulses and a diffraction limited spot size of 21 μm, a strong dependence of the plasma occurrence probability and its intensity on the peak power density was observed. At near threshold values of the peak power density at all pulse repetition rates, the plasma intensity decreased significantly and could not be observed by the CCD camera. However, no systematic dependence between the breakdown threshold and the pulse repetition rate was observed.

Channels were machined in steel and quartz at different values of peak power densities, laser beam traverse speeds and pulse repetition rates. At a peak power density level corresponding to 100 % plasma occurrence probability, significant recast layer was observed in the micro-machined channels. Irrespective of the pulse repetition rate, recast layer was also observed at very low laser beam traverse speeds, such as 1 μm/s.

The best quality surface features with a minimal recast layer, uniform depth and vertical walls was observed when channels were machined in multiple passes at peak power density levels corresponding to 50% - 70% of plasma occurrence probability and laser beam traverse speeds corresponding to a few 100 μm/s. The channel depth varied significantly with the variation in the laser beam traverse speed. At lower values of the laser beam traverse speed, e.g., 100 μm/s, the depth was approximately 30 μm at all pulse repetition rates. However, with an increase in the laser beam traverse speed to values such as 500 μm/s, the channel depth was found to significantly decrease.

4 Conclusion

The feasibility of a new micro-machining process called Laser-Induced Plasma Micro-Machining has been demonstrated. The limitations of μ-EDM and of conventional ultra-short laser micro-machining are not observed with the LIP-MM process. Hence, the LIP-MM process attempts to provide a plausible alternative to the many existing limitations associated with both the μ-EDM and ultra-short laser micro-machining processes.

The LIP-MM process is a very novel process for material removal at the micro-scale. It removes material by plasma-matter interaction. As a matter of fact, it can be postulated that the material removal regime in LIP-MM is very similar to that in μ-EDM. The LIP-MM process also helps in reducing the costs associated with electrode manufacture and dressing in μ-EDM. Finally, it also has the capability to micro-machine materials ranging from quartz which is non-conductive and transparent to metals such as steel, nickel, aluminum, etc.

References

[1] Pham, D.T., Dimov, S.S., Bigot, S., Ivanov, A., Popov, K.: Micro-EDM–Recent Developments and Research Issues. Journal of Materials Processing Technology 149, 50–57 (2004)
[2] Ho, K.H., Newman, S.T.: State of the Art Electrical Discharge Machining (EDM). International Journal of Machine Tools and Manufacture 43, 1287–1300 (2003)
[3] Yu, Z.Y., Kozak, J., Rajurkar, K.P.: Modeling and Simulation of Micro EDM Process. CIRP Annals - Manufacturing Technology 52, 143–146 (2003)
[4] Momma, C., Nolte, S., Chichkov, B.N., Alvensleben, F.v., Tünnermann, A.: Precise Laser Ablation with Ultrashort Pulses. Applied Surface Science 109-110, 15–19 (1997)
[5] Liu, X., Du, D., Mourou, G.: Laser Ablation and Micromachining with Ultrashort Laser Pulses. IEEE Journal of Quantum Electronics 33, 1706–1716 (1997)
[6] Noack, J., Vogel, A.: Laser-Induced Plasma Formation in Water at Nanosecond to Femtosecond Time Scales: Calculation of Thresholds, Absorption Coefficients, and Energy Density. IEEE Journal of Quantum Electronics 35, 1156–1167 (1999)
[7] Shirk, M.D., Molian, P.A.: A Review of Ultrashort Pulsed Laser Ablation of Materials. Journal of Laser Applications 10, 18–28 (1998)
[8] Sacchi, C.A.: Laser-Induced Electric Breakdown in Water. Journal of the Optical Society of America B-Optical Physics 8, 337–345 (1991)
[9] Vogel, A., Nahen, K., Theisen, D., Noack, J.: Plasma Formation in Water by Picosecond and Nanosecond Nd:YAG laser Pulses. I. Optical Breakdown at Threshold and Superthreshold Irradiance. IEEE Journal of Selected Topics in Quantum Electronics 2, 847–860 (1996)

Focused Ion Beam Micro Machining and Micro Assembly

Hongyi Yang and Svetan Rachev

Precision Manufacturing Centre, University of Nottingham, Nottingham

Abstract. The ability to manufacture and manipulate components at the micro-scale is critical to the development of micro systems. This paper presents the technique to manipulate micro/nano parts at the micro/nano-scale using an integrated Focused Ion Beam (FIB) system composed of scanning electron microscope, micro manipulator and gas injection system. Currently the smallest gears manufactured with traditional techniques were reported to have a module of 10 μm. As a test example, in this research we applied the above integrated FIB technique and had successfully fabricated micro gears with module of 0.3μm, with the precision improved for more than thirty times compared to the traditionally manufactured gears. Currently the integrated FIB technique is extended to cover the micro-assembly of devices in our centre, besides the micro-manufacture procedure.

1 Introduction

Focused Ion Beam (FIB) machining is an ideal technique for the fabrication of nano-structured devices with sub-micro scale details; it can also be applied to mill materials that are difficult to deal with other manufacturing methods. For this reason, FIB has found wide applications in many areas, such as MEMS, optical, micro fluidic devices and so on [1]. Recently, application of FIB system has seen ever-increasing applications in micro machining and micro assembly, through integration with other powerful micro/nano facilities such as scanning electron microscopy (SEM), gas injection system (GIS) and nano-manipulator systems.

In this article, we introduce the fundamentals of a newly developed CrossBeam® system, as well as some unique capabilities of the FIB system in micro/nano machining and assembly. The FIB-based micro machining and assembly technique is a promising method for the fabrication and the assembly of complex micro-mechanical systems.

2 Methodology

2.1 NVision 40 Crossbeam® System

In the precision manufacturing centre of Nottingham University, NVision 40 Cross-Beam®, an integrated system combining the FIB and GIS technologies, has been put into use. Figure 1 shows the system schematics. Combined with the core technologies of GEMINI® electron beam technology, the system is the latest member of the industry proven CrossBeam® family of FIB workstations. The system also benefits from a micromanipulator and a 6-axis super eccentric specimen stage. Table 1 illustrates the capabilities and specifications of the NVison 40 system.

S. Ratchev (Ed.): IPAS 2010, IFIP AICT 315, pp. 81–86, 2010.

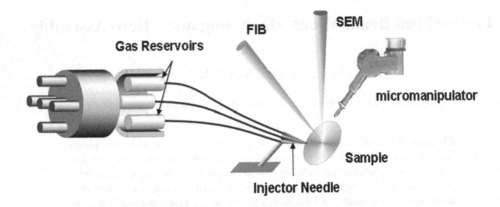

Fig. 1. Schematic of the NVision 40 CrossBeam® system

Table 1. Capabilities and specifications of the NVision 40 CrossBeam®

Capabilities	Specification	Function
SEM (Carl Zeiss SMT)	SEM Resolution: 1.1 nm at 20 kV; 2.5 nm at 1 Kv	Imaging
FIB (SII NanoTechnology Inc.)	FIB Resolution : 4 nm at 30kV	Milling and Deposition
GIS	Four channel single injector GIS (Carbon, platinum, insulator, water, XeF_2)	Enhanced Milling and Deposition
Micromanipulator (Kleindiek nanotechnik, Germany)	Motion Resolution: 5 nm in vertical direction; 3.5 nm in lateral direction; 0.25 nm in back/ forth direction	Manipulation

The FIB imaging can be used for positioning the sample, but the FIB imaging has less contrast and might damage the sample surface, which needs to be taken care of during operation. In the CrossBeam operation mode both beams are turned on, and while the ion beam is milling a defined area, the SEM is used for the real-time capturing of the image of the milling process at high resolution. This enables the operator to control the milling process on a nano-scale, and thus facilitates fabricating extremely accurate cross sections and making necessary device modifications.

2.2 Micro Machining and Micro Manipulation

With the sputtering capability, the FIB can be used as a micro-machining tool to modify or machine materials at micro- and nano-scale accuracy. Instead of milling, materials like carbon, platinum and insulator can be deposited onto the sample surface when precursor gas is injected under the ion beam through the GIS system. Complex geometries can be fabricated by direct milling or deposition in a single process. During the milling/deposition process, the ion beam is applied spot-by-spot under digital control based on the geometry of the designed structure, without using masks as in conventional methods. During the feature machining and deposition process, first the feature geometry is prepared using CAD software or by importing the feature file in BMP or DXF format. Next the NVison control software converts the feature (in CAD or bitmap file format) to raster points, and then controls the beam position according to the raster points with suitable specified dwell time and current. Several factors affect the FIB system, and the optimisation of these factors has been investigated in a previous paper [2].

In this study, a slice sample was cut from the substrate and fixed on a holder. This cutting and deposition process with a micro manipulator is a typical process for the TEM sample fabrication.

To evaluate the capability of micro-manufacturing and assembly, two micro gears are fabricated as a test case study. Both the two micro gears have 15 teeth, while their nominal diameters of the pitch circle are 4.5μm or 13.5μm, and the corresponding modules are 0.3μm or 0.9μm.

3 Results and Discussion

With all the necessary capabilities integrated in one chamber, processes such as cutting, deposition and manipulation can be easily carried out with the FIB system, and the fabrication procedure can be observed in real-time through the integrated SEM capability. Figure 2 shows a TEM sample being cut and lifted by the micro-manipulator and attached to the holder through the deposition process.

Micro parts or micro structure on the slab can be machined with the FIB milling in the designed profile and position. The FIB deposition glues the probe with the micro part, and the probe moves the part with high resolution to the holder or a prepared position. The part is fixed to the holder by deposition or put into a fixture without glue. The probe is then detached from the parts. Using a micro gripper, the process to glue the micro part to the probe can be avoided, thus minimising the pollution to the sample.

Compared with macro assembly, the critical challenges in the micro-assembly include the required positional precision and accuracy, the mechanics of interactions due to scaling effects, and the highly precise hand-eye coordination of the operator [3]. In conventional microscope, there is a bottle neck that limits the observation and depositing of objects with higher precision. With the FIB system, these problems can be improved with the real-time SEM high resolution imaging as well as high resolution manipulation. In the micro scale, releasing the parts becomes difficult because of the mutual forces among the parts, which are often greater than gravitational force. The deposition/cutting procedure is more advantageous over the micro gripper for carrying and releasing micro parts in this case, despite the longer operation time involved.

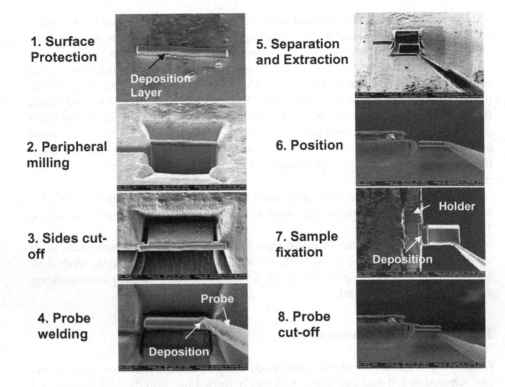

Fig. 2. Process of preparation of a TEM sample

Miniaturization of gears is the key to the development of micro devices like micromotor, which often have outside diameters of a few millimetres. To cater for the future demand for micro-gear, more and more new manufacturing methods and mass production technology are emerging. Current micro gear manufacturing methods include: hobs hobbing processing; micro wire EDM, injection molding of plastic gears; metal sintering; and other semiconductor manufacturing method like lithography or laser processing method. Ogasawara Corporation has achieved the smallest modulus of 0.01mm in micro-gears with the hob cutting technique [4]. However, gears with modulus smaller than 0.01mm is still at the exploratory stage.

In the current study, by fully utilising the advanced NVision 40 system we worked on fabrication of micro-gears with modulus as small as 0.3μm. The superior system capability in precision manufacturing ensured the successful fabrication. Figure 3 (a) and (b) show the top view and lateral view of the fabricated micro-gear with 0.9 μm modulus, and Figure 4 (a) and (b) illustrate the micro-gear with 0.3 μm modulus. Currently we are working on the further development of micro-assembly procedure, to combine the fabricated micro-gears and form a practical gear motion chain. Figure 5 presents the schematic of the planed gear chain.

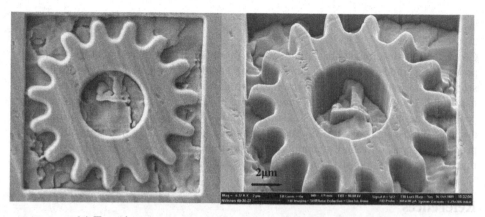

(a) Top view   (b) Lateral view

Fig. 3. Fabricated micro gear with module of 0.9 μm

(a) Top view   (b) Lateral view

Fig. 4. Fabricated micro gear with module of 0.3 μm

Fig. 5. Planning of micro gear assembly

4 Conclusion

CrossBeam® system provides an integrated manufacturing environment for the rapid-prototyping, modification and fabrication of small series of nano structures in one-stop and the manipulation after the machining, which can find wide applications in nanoscience and nanotechnology practice. We have developed a micro-manufacturing procedure with superior micro-assembly capability by employing FIB milling, FIB assisted deposition, and the micro manipulation system. This micro-manufacturing procedure is a powerful technique for the fabrication and assembly of micro-machines, as validated by our test fabrications so far.

References

[1] Watt, F., Bettiol, A.A., Van Kan, J.A., et al.: Ion beam lithography and nanofabrication: a review. Int. J. Nanosci. 4, 269–286 (2005)

[2] Yang, H., Ronaldo, R., Segal, J., Ratchev, S.: Focused ion beam nano machinining, Lamdmap 9th, London (2009)

[3] Tietje, C., Leach, R., Turitto, M., Ronaldo, R., Ratchev, S.: Application of a DFμA methodology to facilitate the assembly of a micro/nano measurement device. In: Ratchev, S., Koelemeijer, S. (eds.) Micro-assembly technologies and applications. IFIP, vol. 260, pp. 5–12. Springer, Boston (2008)

[4] http://www.djwxw.com/News/HtmlPage/2006-02-16/TT_1206_1.htm

Chapter 4

Innovative Assembly Processes

Hybrid Assembly for Ultra-Precise Manufacturing

Alexander Steinecker

Centre Suisse d'Electronique et de Microtechnique S.A.,
Central Switzerland Center,
Untere Gründlistrasse 1,
6055 Alpnach-Dorf, Switzerland

Abstract. This paper presents a European integrated project in the domain of innovative hybrid assembly - the combination of self-assembly and robotics. Robotics allows for high precision and flexibility with full control of the assembly process. Self-assembly, on the other hand, offers massively parallel, unsupervised processing and thus provides high throughputs that cannot be reached with simple robotics alone. The combination of these approaches thus offers robust, high-speed and high-precision assembly methodologies.

Different approaches to achieve reliable self-assembly are investigated, such as chemical or physical surface modification and structuring and field induced assembly. These techniques are combined with high performance robotic tools such as precise manipulators, innovative vision, force sensing and system control. Selected innovative demonstrator systems prove the industrial impact of these developments. Focus is put on assembly and processing of small parts with dimensions below 1 mm: MEMS parts, RFID tags, cells, optical systems and nanowires.

Keywords: Hybrid assembly, robotics and self-assembly, new production paradigm.

1 Introduction

1.1 Core Approach

Today, emerging highly complex micro-devices with applications in mechanics, electronics, biological engineering, microfluidics and IT demand ultra precision and flexible assembly [1], [2], [3]. A need becomes apparent to develop tomorrow's manufacturing processes for complex micro-products.

Hybrid assembly has been identified as a promising, innovative production technology. It is defined as the combination of two approaches: (i) *positional robotic assembly* where objects are mechanically manipulated and positioned one by one; and (ii) *self-assembly* where objects arrange themselves into ordered structures through physical or chemical interactions. Such hybrid assembly has not yet been achieved at the industrial scale. It will build a production paradigm permitting the development and implementation of fully innovative production processes for assembly of a variety of microproducts.

S. Ratchev (Ed.): IPAS 2010, IFIP AICT 315, pp. 89–96, 2010.

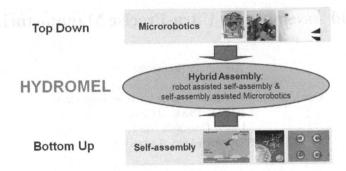

Fig. 1. The HYDROMEL project aims at combining positional robotic assembly with self-assembly to develop a hybrid production approach

1.2 The HYDROMEL Project

An integrated European project has been launched in order to address the ambitious goal of implementing hybrid assembly at the industrial scale. The HYDROMEL project has been implemented to establish hybrid assembly as an impending production paradigm. The project which began in October 2006 will continue for 4 years concluding in September 2010. It brings together experts in the robotics and self-assembly fields from academia and R&D as well as high-level industrial partners. Currently 23 parties from 9 European countries are involved. Selected industrial partners (highlighted in Fig. 2) will act as end-users demonstrating various applications generated by the project.

The industrial exploitation and implementation is assured by core partners inside the consortium and supported by an external highly qualified industrial advisory board.

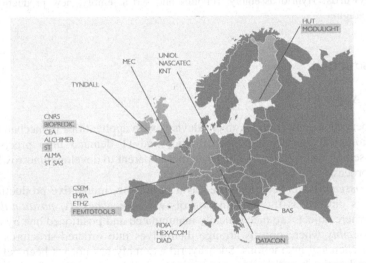

Fig. 2. Partners that are involved in the HYDROMEL project. Selected industrial partners demonstrating application cases are highlighted.

1.3 R & D Progress

R&D activities focus on the following domains: (i) *microrobotics*; (ii) *self-assembly;* and (iii) the combination of both microrobotics and self-assembly to form *hybrid assembly.*

Two hybridization scenarios will be studied. The first scenario focuses on *robotics assisted by self-assembly* in which a robotic process is supported and optimized through self-assembly. For example, the use of self-directing capillary effects and low-adhesion functionalized surfaces will be studied for reliable part handling. The complementary scenario will also be studied in which a self-assembly or *self-alignment process is supported by robotics.* The massively parallel, unassisted positioning of parts (e.g. nanowires) on a substrate and the complementary use of robots to individually correct or manipulate specific target objects is one example of this approach. Another example of this second scenario is the combination of coarse robotics and precise self-assembly. Here the robot initiates the self-assembly process by roughly positioning the part near the attractor allowing the self-assembly process to complete the precise positioning of the part.

Hybrid assembly is aimed to be introduced for a broad range of applications on the industrial scale with very different demands in terms of component size, accuracy and throughput. In the following, five scenarios are briefly sketched: (i) hybrid assembly of fragile MEMS parts; (ii) hybrid assembly for high-throughput production of electrical devices; (iii) self-alignment assisted manipulation of biological cells; (iv) hybrid assembly of nano-structures such as nanowires or nanotubes for photonic applications; and (v) self-alignment in high-throughput quality inspection.

2 The HYDROMEL Technology Platform

Within the project, R&D activities in robotics, self-assembly and their combination is carried out. These activities result in a technology platform that serves as a toolbox to set up different hybrid assembly applications. These applications are demonstrated in selected industrial systems.

2.1 Robotics

Robotics is a well established industrial technology. Nevertheless, if high throughput and high accuracy are desired - especially in the microrobotics area - several needs have been identified: (i) High-speed nano-manipulators by combining long range fast conventional robots and precise nano-robots; (ii) tools and handling strategies for micro-handling; and (iii) control and sensor fusion for process control of microhandling.

Improved components for reliable microhandling have been developed. Selected highlights of these improvements are the combination of nanorobots and innovative multidimensional vision sensing for closed-loop robot operation at nanometer precision (in the scanning electron microscope), and the development of tools (gripper, feeder) for microcomponents.

2.2 Self-assembly

Self-assembly can be considered as a new strategy for nano- and microfabrication. It offers a bottom-up production technique with massive parallel throughput. Self-assembly methods based on programmable forces for self-assembly of a range of mesoscale (length scale of parts in the micro- to millimeter-range) components have been developed including the following approaches: (i) surface treatment and patterning for different self-assembly objects – biological cells, chips, and nanowires; (ii) chemical and physical switching of surface properties for controlled adhesion; (iii) directed self-assembly by application of external fields; and (iv) modeling and measuring of bonding forces in self-alignment.

Important goals in the development of self-assembly technologies have been achieved in the HYDROMEL project. The feasibility of various techniques (local surface treatment, global switching of surface properties) could be proven for selected model systems. Precise self-assembly of chips on a substrate has been realized (Fig. 3). Furthermore, important progress has been made in field-induced self-assembly of nanowires.

Fig. 3. High-precision fluidic self-assembly (courtesy of HUT)

2.3 Hybrid Assembly

Hybrid assembly is the core technology bringing together robotics and self-assembly. Two ways to implement hybrid assembly are investigated. Several examples of hybrid assembly scenarios are given below:

Improvement of robotics by using self-assembly techniques
Classical robotics can be improved by means of self-assembly in different ways. Structuring techniques can be applied to grippers in order to improve picking, reliable positioning and releasing of micro-objects. Feeding - a classical robotics task - can also benefit from controlled and switchable self-assembly by collecting objects in desired position and thus facilitate the efficient pick- and place process.

Improvement of self-assembly by using robotics
Using robotics to assist or improve self-assembly is useful in various applications. Micro- or nanorobots can be used in processes where the result of a self-assembly process has to be corrected or characterized. Robust and fast coarse robotics will be used to place meso-scale objects close to self-alignment attractors (an example of a

mechanical self-alignment structure is given in the Fig. 4). The final alignment in position and/or orientation that reaches the targeted accuracy is carried out by unsupervised self-alignment. As a result, parallelization can be achieved as well as a reduction of investment of equipment.

The main project activity is the implementation of hybrid assembly and integration into dedicated systems. The functionality of modules and sub-systems has been proven for various applications with industrial relevance.

Fig. 4. Clipping structures that allow precise fiber positioning with coarse robotic approach (courtesy of CSEM)

3 Technology Demonstrators for Industrial Applications

The industrial integration of the core technologies that have been presented in the previous section is demonstrated for selected applications. The technology demonstrators address complementary assembly cases with very different system specifications and application areas.

3.1 Advanced Micromechanics: Hybrid Assembly of Fragile MEMS Parts

MEMS parts for measuring micro forces and for force controlled microgripping must be assembled into a package. It is important to obtain a mechanical and electrical connection between the fragile MEMS component and a printed circuit board (PCB). The MEMS parts consist of fine mechanical structures and are very brittle. A parallel handling process is implemented with the aim of overcoming tedious one-by-one processing. Coarse robotics will be combined with precise self-alignment to achieve high throughput and high alignment accuracy of the package.

Fig. 5. Force sensor. System overview (left) and close-up view (courtesey of FemtoTools).

The production equipment will be cost effective, reduce assembly effort and increase the process yield targeting mid- to high-production rates.

3.2 Electronics: Hybrid Assembly of RFID Tags

A hybrid assembly solution for high-precision and high-speed assembly of RFID chips on an antenna web will be developed. The state-of art production process for RFID tags is a pick-and-place procedure with dedicated and highly optimized die bonding equipment. An alternative industry compatible approach will be implemented offering the opportunity for parallelization and cost reduction by an optimum combination of coarse robotic placement and fine capillary self-alignment of the RFID chip. High-speed and high-throughput processes are targeted.

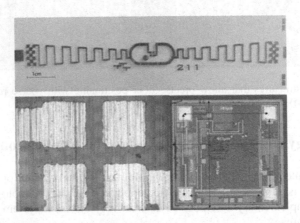

Fig. 6. Antenna web (top) and RFID chip with contact pads (bottom, courtsey of DATACON)

3.3 Bio- and Life-Sciences: Self-alignment Assisted Handling of Cells

A lab-automation system is being integrated that combines several aspects of cell handling. It offers automated procedures for cell selection, immobilization, and a microinjection process. The system combines microfluidics for prior cell sorting and separation, self-assembly for reversible immobilization of the cell, and microrobotics for force-controlled cell injection. The complete system will be an automated, high throughput cell processing system. Specifically, cells with sizes between 1 and 0.02 mm will be targeted.

3.4 Future Technologies: Self-assembly for Emerging Nanophotonics and Nanoelectronics

A palette of hybrid robotic / self-assembly techniques for nanowires or nanotubes will be developed and applied to a variety of nanostructures. The assembly methodology will be demonstrated but also the function of resulting nanoscale devices. Self-assembly of nano-objects will allow the organization of nanoscopic objects much smaller than those that can be defined with classical top-down approaches. This approach is mandatory for Systems on Chip or even in future 3D circuits.

Fig. 7. System for hybrid cell-sorting and micro-injection (courtesey of CSEM)

Hybrid approaches are investigated addressing production and characterization of self-assembled nanostructures combined with robotic quality control and error-correction measures. Key technologies are: field-induced, guided assembly of nanowires; self-alignment of large areas of nanowires; and, individual robotic manipulation of nano-components for selective processing.

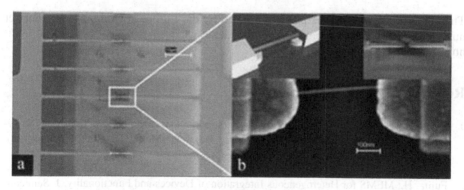

Fig. 8. Hybrid nanofabrication approach to realize integrated nano-systems (a) Nanoarray design, (b) multi-walled nanotube on a nanostructure with schematic and a scanning electron microscope image shown in insets (courtesy of ETHZ)

3.5 Opto-electronics: Hybrid Self-alignment in Optical Inspection

A high-speed handling and inspection solution for laser diodes has been developed which overcomes precise pure robotic pick-and-place approaches. The hybrid inspection system handles laser diode chips with sizes down to 0.1 mm. The system combines coarse handling robotics with high precision self-alignment and self-positioning of the chips.

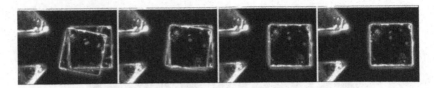

Fig. 9. Study of dynamic self-alignment of microparts(courtesy of HUT)

4 Outlook

Hybrid assembly defined as combination of self-assembly and robotics offers a new assembly paradigm in various production scenarios. It allows for high-precision and high-throughput assembly processes for hybrid micro-devices. An essential component of successful implementation is the application of specific design rules already in an early product development stage, taking into account integration of features that support self-assembly (e.g. surface structure / chemical composition of target areas).

Conventional robotics can benefit from the self-positioning and -alignment of components in target positions. On the other side, high specificity and customization of self-assembly can be achieved by combining them with advanced robotics.

Acknowledgements

The HYDROMEL project is partially funded by the European Commission, contract number 026622-2. The support is gratefully acknowledged. See also http://www.hydromel-project.eu.

References

1. Böhringer, K.F.: Heterogeneous System Integration by Self-assembly. Talk given at the hybrid and self-assembly workshop, University Brussels (March 12, 2009) (manuscript), http://beams.ulb.ac.be/beams/perso/plambert/hydromelworkshop/bohringer.pdf
2. Fujita, H.: MEMS for Heterogeneous Integration of Devices and Functionality. J. Semiconductor Technology and Science 7(3), 133–138 (2007)
3. Saeedi, E., Draghi, L., Parviz, B.A.: Optically Programmable Self-Assembly of Heterogeneous Micro-Components on Unconventional Substrates. In: Proceedings of MEMS 2009. IEEE 22nd International Conference, January 25-29, pp. 717–720 (2009)

Development of the Roll Type Incremental Micro Pattern Imprint System for Large Area Pattern Replication

Jung-Han Song[1], Hye-Jin Lee[1], Shuhuai Lan[1], Nak-Kyu Lee[1], Geun-An Lee[1],
Tae-Jin Lee[1], Seogou Choi[1], and Sung-Min Bae[2]

[1] Manufacturing Convergence R&D Department, Korea Institute of Industrial Technology,
1271-18 Sa 3-dong, Sangrok-gu, Ansan-si, Gyeonggi-do, 426-173 Korea
[2] Dept. of Industrial & Management Engineering, Hanbat National University Korea
{Jhsong,naltl,bluetree,nklee,galee,ltj0822,schoi}@kitech.re.kr,
loveiris@hanbat.ac.kr

Abstract. Flexible display has been attracting attention in the research field of next generation display in recent years. And polymer is a candidate material for flexible displays as it takes advantages including transparency, light weight, flexibility and so on. Rolling process is suitable and competitive process for the high throughput of flexible substrate such as polymer. In this paper, we developed a prototype of roll-to-flat (R2F) thermal imprint system for large area micro pattern replication process, which is one the key process in the fabrication of flexible displays. Tests were conducted to evaluate the system feasibility and process parameters effect, such as flat mold temperature, loading pressure and rolling speed. 100 mm × 100 mm stainless steel flat mold and commercially available polycarbonate sheets were used for tests and results showed that the developed R2F system is suitable for fabrication of various micro devices with micro pattern replication on large area.

Keywords: Flexible Display, Roll-to-Flat Thermal Imprint, Micro Pattern Replication, Polycarbonate.

1 Introduction and Background

As the semiconductor industry and IC (Integrated Circuit) industry developing and flat panel display market growing there are demands to search a method that can provide production with high precision and good quality at a low-cost and high-throughput way. Among them, Micro- or Nano- structure fabrication or high-precision nanoscale lithography process is the key technology to the manufacturing of photonic components, micro - and nanofluidic chips, chip-based sensors and biological applications. Most manufacturer and researcher are looking for a proper fabrication way. Except for the conventional technologies such as well-known Optical lithography and Electronic Beam lithography, several alternative approaches towards nanostructure fabrication have been exploited, for example, Micro-contact Printing, Nanoimprint Lithography, Scanning-Probe-Based techniques, Dip-Pen Lithography and Nanoplotting, as well as Stenciling. [1-3]. Among them Nanoimprint lithography

S. Ratchev (Ed.): IPAS 2010, IFIP AICT 315, pp. 97–104, 2010.

(NIL) was considered widely as the potential high-throughput, high-resolution and low-cost manufacturing method. [4]

Early in 1995, S.Y. Chou et al. firstly proposed the first Nanoimprint Lithography process in NanoStructure Laboratory of University of Minnesota. They firstly realized patterning sub-25 nm and sub-10 nm nanostructures in the polymers on a silicon substrate, which is followed by Reactive Ion Etching (RIE) or dry etching process. During the process, a pattern mold was pressed onto a thin thermoplastic polymer film on a substrate to create vias and trenches with a minimum size of 25 nm and a depth of 100 nm in the polymer. [4-5]

The research on the NIL has spread over the world and distinguished results have been achieved. By now, they had generated many kinds of NIL technologies. Generally speaking we can classify them into three categories of technology: (1) Hot-Embossing Nanoimprint Lithography (HE-NIL) [4,5] (2) UV-type Nanoimprint Lithography (UV-NIL) [6-8] (3) Soft-Lithography [9]. Now the three kinds of NIL technologies have developed more and have been used widely in many different applications, such as biomedical product [10], data storage [11], optical parts [12], organic electronics [13] and so on. However, there are still many problems should be solved in the conventional NIL process so that it can be used as mass production method in the industry. One of the most important problem is that it cannot significantly improve the throughput in the patterning of large area product with low cost because it is not a continuous process.

To overcome this problem Roller-type Nanoimprint Lithography (RNIL) [14] was firstly proposed by S.F. Chou et al. and has developed quickly and become the potential manufacturing method for industrialization of nanoimprinting process, due to its prominent advantage of continuous process, simple system construction, high-throughput, low-cost and low energy consuming. For the conventional flat nanoimprint, the whole area of the sample is imprinted at the same time. But for RNIL, only the adjacence area of contact line is pressed for a given time, significantly reducing the effects of thickness unevenness and dust from environment. So in this article, our aim is to give a brief review on the state-of-the-art of the novel NIL process.

2 Continuous Roller-Type Nanoimprint Lithography Process

Actually Roller-type manufacturing (Roll to Roll) methods has previously used in many industrial fields, such as gravure printing or flexography printing (or flexo). It was used to imprint the pattern of the roller on flexible thin films. They were traditionally used for printing newspapers, magazines and packages. Especially in the field of manufacturing for flexible electronics, this fabrication method has already developed. Flexible electrophoretic displays manufactured by roll-to-roll (R2R) processes have been demonstrated good reliability and performance. [15] Wang Xiaojia et al. have also developed a novel full color electrophoretic film manufacturing process using Roller-type method. [16] In the actual industrial manufacturing, typical R2R manufacturing consists of three essential steps, which are deposition, patterning and packaging, as shown in Fig. 1. [17-18].

As its name implying, Roller-type manufacturing method has the advantage of providing a continuous and high throughput. When it was novelly used in the nano-imprint lithography technology, the cost of NIL production can be reduced.

Early in 1998, Hua Tan et al. firstly proposed and demonstrated an alternative approach to the NIL process, the Roller-type Nanoimprint Lithography process. In this process, they had achieved some sub-100 nm resolution patterns transfer. [14] The schematic of this method is shown as Fig. 2. Here the roller-type nanoimprint lithography process system mainly consists of three function components: a roller, a movable platform, and a hinge. They developed two methods for RNIL: (a) rolling a cylindrical mold on a flat and solid substrate; (b) putting a flat mold directly on a substrate and rolling a smooth roller on the back of the mold as show in fig. 4. In the first method, the cylindrical mold has thin metal film mold around a smooth roller. On the thin metal sheet, micro patterns were formed For example the master mold made of 100 µm thick of Ni with 700-nm - wide tracks was used for a compact disk. In both methods, the roller temperature was set above the glass transition temperature Tg of the resist, PMMA (Polymethyl Methacrylate) in Fig. 3, while the temperature of the platform is set below Tg. Therefore, only the area in contact with the roller has a temperature higher than Tg, This makes the resist in the area molten and imprinted with the patterns. This is different from flat nanoimprint, where the entire resist, heated above Tg, was imprinted simultaneously and the pressure is applied until the resist is cooled down and cured.

Evolving these two methods, many kinds of Roller-type imprint lithography process have been developed We categorize the Roller-type NIL as methods, using roller mold and using flat mold. In the following, some typical and distinguished research will be introduced.

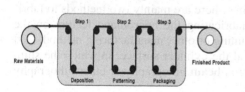 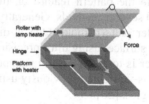

Fig. 1. Three steps in industrial R2R process   **Fig. 2.** Schematic of a roller nanoimprint system

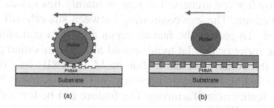

Fig. 3. Two methods of RNIL: a) using roller mold; b) using flat mold & smooth roller

2.1   RNIL Process Using Roller Mold

A roller pattern mold is widely used in RNIL. It can be used both in thermal type (Hot-Embossing) and UV type (UV-based) RNIL process. In the thermal type RNIL process, by rotating and pressing the cylinder mold into the polymer film coated on the substrate with the temperature over glass transition (Tg), the feature on the mold is transferred to the film and the film is pushed backward at the same time. By cooling down, the transferred pattern is cured. While in the UV type RNIL process, the UV light is illuminated when the rotating roller mold contacts with the coated film. The original phase of film is gel or liquid. By illuminating the light, the phase chenged to solid and then the feature on the roller mold is replicated as shown in fig. 4.

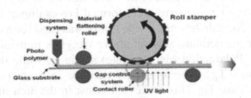

Fig. 4. Schematic diagrams of the UV-type RNIL system for rigid substrate

No matter in Thermal type or UV type RNIL process, the fabrication of roller mold is important, as it directly involves the replication quality of the imprinting process. The manufacturing accuracy of the mold decides the resolution of the RNIL product. And it also acts as a barrier of the development of the RNIL technology. As the target feature size gets smaller below several tens of micrometers, it is difficult to fabricate. To create the pattern feature on the surface of the mold cylinder, researcher have searched several methods. Generally speaking, there are mainly two methods to fabricate roller pattern mold. One is directly fabricating the micro/nano feature on the cylinder surface, using ultra-precision machining or other manufacturing method. [19] The other is to cover a shim stamper around the cylinder surface, in which the shim stamper is fabricated previously using electron beam lithography or other lithography technologies. [20]

2.2   RNIL Process Using Flat Mold and Smooth Roll (R2F)

Comparing with RNIL using roller mold, the RNIL using flat mold (R2F Process) is only being studied by few researchers. The reason mainly lies in that it cannot realize the continuous replication. The gap controlling between smooth roll and flat mold is additionally required. To pattern the feature repeatedly as a continuous process, the transportation of the roller or the flat mold should also be embodied precisely as well as the rotation of the roller. But using flat mold in RNIL has its advantage on mold fabrication, including fabrication accuracy and cost, as it has been at a mature level of micro/nano feature manufacturing. The feature can be formed using the fabrication methods such as MEMS, ultra-precision machining, electron beam maching or photolithography.

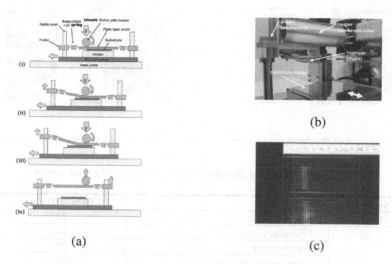

Fig. 5. (a) RNIL system concept using flat mold and smooth roller from AIST; (b) Picture of the real equipment; (c) Imprinting result on a flexible substrate

Several research group proposed concepts using smooth roller and flat mold in their process. AIST(Advanced Industrial Science and Technology) in Japan proposed a process for large area replication. The configuration is shown as Fig. 5. Micro/nano feature on the mold can be replicated to a polymer substrate by applying force and heat with a heater combined in the roller. An automatic mold releasing mechanism is installed in the system to avoid transferred pattern deformation error by mismatch of thermal expansion ratio between a mold (metal) and a substrate (polymer). The distinguished advantage is combining the smooth roller and a movable stage that contains the substrate and a flexible plate mold. By pressing the smooth roller to the flexible plate mold, the pattern can be transferred to the substrate, at the same time the rotation of the roller push the movable stage backward. During the process, smooth roller and substrate were heated by their heater elements respectively. Good result was claimed in their report, as shown in the Fig. 5. (c).

3   R2F Imprint System Development

In this study, we developed a prototype of lab-scale thermal RNIL equipment using flat mold (R2F System) based on the survey. A photograph of the R2F system is shown as Fig. 6. It is made up of several major sub-units, including roller imprint unit, polymer holding and auto-releasing unit, control unit and PC unit for process parameters input (temperature, speed, force, etc.) and result data acquisition.

Fig. 7 shows the photographs of the R2F micro thermal imprint system consisting of major components such as the pressing roller with heaters inserted into the core, which is actuated by AC-servo motor, the movable stage, flat mold combined with a heater, and polymer substrate holding and auto-releasing device. The polymer substrate supply and holding device is made up of a pair of linear grippers and a pair

Fig. 6. Overview of the developed system   **Fig. 7.** Photographs of major components

Table 1. Specifications of the roll-to-flat micro thermal imprint system

System size (mm)		1500×1100×1800
Stage stroke (mm)		700
Roller	Diameter (mm)	160
	Width (mm)	290
	Maximum temperature (℃)	200
	Maximum loading pressure (tonf)	1
	Scanning speed (mm/s)	0.1-10
	Material	Stainless steel
Flat Mold	Maximum mold area (mm²)	150×150
	Maximum temperature (℃)	300
	Moving speed (mm/s)	0.1-10

of tension rollers. This device can adjust the small gap between polymer substrate and flat mold. When the pattern replication is being performed in the semi-linear contact area, other area of polymer will not contact with the mold. After rolling and the mold moving forward, the replication area completed will automatically be separated from the mold. The concept is introduced because the imprinted pattern can be released vertically without damage from the flat mold since the continuous rolling movement.

The movable stage, supporting the flat mold, can move parallel to the sample surface in one direction, at the speed range from 0.1 mm/s to 10 mm/s. The maximum distance allowed for the movable stage is about 500 mm. The maximum press force for the pressing rolling device is 500 kgf with a precision of less than 1 N. The detailed specification of the system is summarized in Table 1.

A stainless steel flat mold with size of 100 mm × 100 mm was used in the experiments, which was fabricated by dicing process. The entire surface of the mold is composed with square micro structure arrays with height of 90 µm, width of 200 µm and 110 µm spacing. Fig. 8 shows a photograph of the flat mold that is embedded on the movable stage. Fig. 9 illustrates a typical 3D scanning image of a mold sub-feature using whitelight scanning interferometry (SNU Precision, Korea) Commercial polycarbonate(PC) substrates were imprinted in the tests. Under the process condition that flat mold temperature is above PC's glass transition temperature (150 ℃) and low rolling speed, the micro patterns on the flat mold were transferred to the PC substrates.

Fig. 8. Photograph of the flat mold prepared   **Fig. 9.** Scanning 3D image of mold sub-feature

4 Result and Discussion

In order to evaluate the performance and feasibility of the developed R2F micro thermal imprint system and to study the rolling process parameters, such as imprinting temperature, loading pressure and rolling speed, several experiments were conducted.

Using the whitelight scanning interferometry the tomograph of the micro pattern on the flat mold and the imprinted pattern on the substrate surface were measured and compared. Based on these measured results, the evaluation of process formability was carried out.

Trial tests were conducted using a rigid and thick PC substrate with thickness of 2 mm. Fig. 10 and 11 shows an example of replication result from flat mold to PC substrate. The process was conducted under the condition of 200 kgf of press force, 1 mm/s rolling speed (also the moving speed of movable stage) and 160 °C temperature of flat mold. The roller was not heated and it was in room temperature. From the result shows that good replication results of uniform micro cup arrays were fabricated over the whole surface area. Some defects on the beginning part (on the bottom of the figure) due to the change of rolling speed from 0 to the constant speed 2 mm/s at the beginning of process.

Fig. 10. Photograph of the imprinted substrate   **Fig. 11.** SEM image of imprinted patterns

5 Conclusion

In summary we have developed a prototype of lab-scale roll-to-flat (R2F) thermal imprint system for large area micro pattern replication process as a potential application of flexible display based on the survey. Series of tests were carried out using a stainless steel flat mold at size of 100 mm × 100 mm and commercial available polycarbonate (PC) substrates to evaluate the system feasibility and process control parameters effect, such as flat mold temperature, loading pressure and rolling speed. The test results show that the developed R2F system is suitable for fabrication of micro devices with micro pattern replication at large area.

Acknowledgement

This work has been financially supported by Korea Research Council for Industrial Science & Technology through the project of Convergence Manufacturing Technology Development of Bio-Medical Applicable Systems for Geriatric Diseases (B551179-09-02-00). The authors are grateful to the colleagues for their essential contribution to this work.

References

1. Jay Guo, L.: Nanoimprint Lithography: Methods and Material Requirements. Advanced Material 19, 495–513 (2007)
2. Gates, B.D., et al.: New Approaches to Nanofabrication: Molding, Printing, and Other Techniques. Chemical Reviews 105, 1171 (2005)
3. Sotomayor Torres, C.M., et al.: Nanoimprint lithography: an alternative nanofabrication approach. Materials Science and Engineering C 23, 23–31 (2003)
4. Chou, S.Y., et al.: Imprint of sub-25 nm vias and trenches in polymers. Applied Physics Letters 67(21), 3114–3116 (1995)
5. Chou, S.Y., Krauss, P.R.: Imprint Lithography with Sub-10 nm Feature Size and High Throughput. Microeleetronic Engineering 35, 237–240 (1997)
6. Bender, M., et al.: Fabrication of nanostructures using a UV-based imprint technique. Microelectronic Engineering 53, 233–236 (2000)
7. Otto, M., et al.: Characterization and application of a UV-based imprint technique. Microelectronic Engineering 57-58, 361–366 (2000)
8. Colburn, M., et al.: Step-and-flash Imprint Lithography: A New Approach to High Resolution Patterning. In: Proc. of SPIE, vol. 3676, p. 379 (1999)
9. Xia, Y., Whitesides, G.M.: Soft Lithography. Angewandte Chemie International Edition, vol. 37, pp. 550–575 (1998)
10. Pepin, A., et al.: Nanoimprint lithography for the fabrication of DNA electrophoresis chips. Microelectronic Engineering 61-62, 927–932 (2002)
11. Nilsson, M., Heidari, B.: Breaking the Limit-Patterned Media for 100 Gbits and beyond. Obducat, Malmo, Sweden
12. Park, Y.K., Kostal, H.: Nano-Oprics Redenfine Rules for Oprical Processing. Communication System Design, 23–26 (August 2002)
13. Clavijo Cedeno, C., et al.: Nanoimprint lithography for organic electronics. Microelectronic Engineering 61-62, 25–31 (2002)
14. Tan, H., Gilbertson, A., Chou, S.Y.: Roller nanoimprint lithography. Journal of Vacuum Science & Technology B 16(6), 3926–3928 (1998)
15. Hou, J., et al.: Reliability and performance of flexible electrophoretic displays by roll-to-roll manufacturing processes. In: SID 2004 Digest, Seattle, USA, pp. 1066–1069 (2004)
16. Xiaojia, W., HongMei, Z., Li, P.: Roll-to-roll manufacturing process for full color electrophoretic film. In: SID 2006 Digest, San Francisco, CA, pp. 1587–1889 (2006)
17. Grawford., G.P., et al.: Roll-to-Roll Manufacturing of Flexible Displays. Flexible Flat Panel Displays, 410–445 (2005)
18. Schwartz, E.: Roll to Roll Processing for Flexible Electronics. Cornell University MSE 542: Flexible Electronics, May 11 (2006)
19. Ahn, S., et al.: Continuous ultraviolet roll nanoimprinting process for replicating large-scale nano- and micropatterns. Applied Physics Letters 89, 213101 (2006)
20. Makela, T., et al.: Continuous roll to roll nanoimprinting of inherently conducting polyaniline. Microelectronic Engineering 84, 877–879 (2007)

Utilisation of FIB/SEM Technology in the Assembly of an Innovative Micro-CMM Probe

Daniel Smale[1], Steve Haley[1], Joel Segal[1], Ronaldo Ronaldo[1], Svetan Ratchev[1], Richard K. Leach[2], and James D. Claverley[2]

[1] Precision Manufacturing Centre, University of Nottingham, UK
[2] Industry & Innovation Division, National Physical Laboratory, Teddington, UK

Abstract. The measurement of features from the micro- and precision manufacturing industries requires low uncertainties and nano-scale resolution. These are best delivered through ultra precise co-ordinate measuring machines (CMMs). However, current CMMs are often restricted by the relatively large and insensitive probes used. This paper focuses on the assembly challenges of a novel micro-CMM probe. The probe is comprised of a 70 µm glass sphere, attached to a solid tungsten-carbide shaft of diameter less than 100 µm, joined to a piezoelectric flexure structure. The assembly requirements are for positional accuracy of ± 0.5 µm, angle between the shaft and flexure of 90° ± 0.29° and that the components be undamaged by the process. A combined Focused Ion Beam and Scanning Electron Microscope machine (FIB/SEM) with integrated nano-resolution manipulators was used. The investigation has evaluated potential assembly and joining solutions, identified modifications to existing equipment and product design and produced a set of prototypes.

Keywords: Ultra Precision Assembly, Microscopy, FIB Machining.

1 Introduction

Miniaturisation and integration of mechanical, sensing, and control functions within confined spaces is becoming an important trend in designing new products for commercial sectors such as medical, automotive, biomedical, consumer electronics, and telecommunications [1]. However, the potential for micro-assembly for such products has been shown mainly in the research environment with a limited transfer of knowledge and equipment to industry [2]. These key commercial sectors require that the related products are manufactured to a very high quality and reliability; therefore, metrology is of great importance [3].

The measurement and quantification of features and products from the microscale and precision manufacturing industries often requires uncertainties and resolution in the nanometre range. The best means of delivering these requirements is through high accuracy CMMs [4]. However, the current state-of-the-art CMMs are often restricted by the relatively large and insensitive probes used. An example of such a machine is the Zeiss F25, shown in Figure 1. The F25 has a quoted volumetric accuracy of 250 nm. Work at the Precision Manufacturing Centre (PMC) and the National Physical

S. Ratchev (Ed.): IPAS 2010, IFIP AICT 315, pp. 105–112, 2010.
© IFIP International Federation for Information Processing 2010

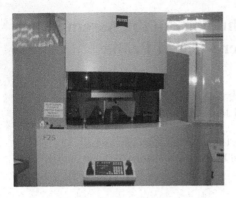

Fig. 1. Image of the Zeiss F25 CMM, located in the PMC facilities

Fig. 2. Image of the micro CMM probe, designed by NPL

Laboratory (NPL) has suggested that the F25 is more accurate than claimed by Zeiss and that the existing probes do not enable full exploitation of the machine's resolution.

The micro-CMM probe discussed in this paper was developed at NPL to help realise the accuracy and traceability required by the microscale and precision manufacturing industries [5]. The probe is comprised of a solid shaft, a flexure assembly, and a probing sphere. A 3D model of the device is shown in Figure 2. The shaft is manufactured from tungsten carbide (WC) via EDM and wire eroding. The flexure assembly is a laminar structure, manufactured by a micromachining process. The flexures include PZT actuators and sensors that are deposited onto the surface of the structure during the manufacturing process. The probing sphere attached to the end of the WC shaft is made of silica. The shaft is 100 µm in diameter where it joins the flexure and 50 µm in diameter where it joins the probing sphere. The shaft is connected at the thick end to the delicate piezoelectric flexure structure via a 50 µm diameter spigot. At the thin end a 70 µm diameter glass sphere is connected concentrically. These joints must be made without damaging any of the components, but special attention is placed on the protection of the sphere. The assembly requirements for the shaft onto the flexure specify a positional accuracy of ± 0.5 µm and the angle between the shaft and flexure to be 90° ± 0.29°. These factors are of primary importance in ensuring correct function of the final product. Assembly is achieved through the application of micro-manipulators integrated within a combined FIB/SEM. The joint is made permanent through two options: adhesive bonding and FIB material deposition.

2 Methodology

2.1 Assembly Challenges

The micro-CMM probe described presents a number of specific assembly challenges. These challenges result from the scale of the parts and the technical requirements of the product. A study by Van Brussel [6] found that as the radius of a part falls below 1 mm,

SEM beam offers imaging with high resolution at magnifications in excess of 300 000x. The FIB is able to machine samples at scales ranging from 10 nm to 10 000 nm. The functionality of the FIB is enhanced by the GIS, which can deliver several materials to be deposited onto the sample with the same resolution. The integrated manipulators have a resolution of 1 nm and are controllable in four degrees of freedom. They have modular tooling, enabling rapid changes of end effectors for different functions.

3 Experimental Work

3.1 Assembly Task 1 Preparatory Work

Assembly Task 1 was the task of joining the shaft to the piezoelectric flexure. This was to be done via a hole in the centre of the flexure. However, the manufacturing process used by NPL could not generate this feature reliably; as shown in Figure 5 the hole was blocked by a film formed during the manufacture of the flexure. Therefore, the initial work for this task was to machine the hole and this was attempted with the Zeiss NVision. The hole was required to be 100 µm in diameter and through the approximately 15 µm thick flexure. As the Zeiss NVision is designed for nano-scale machining, these are comparatively large dimensions.

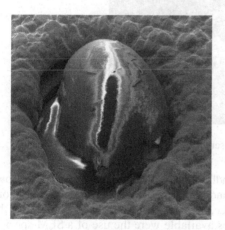 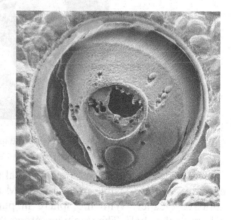

Fig. 5. Obstruction of the hole in the flexure   **Fig. 6.** The obstruction has fallen out of the flexure

The strategy adopted at first to remove the obstruction was to simply machine out the whole area. However, this was unsuccessful; after several hours of machining with high power settings (30 kV and 6.5 nA) the FIB had only begun removing the surface of the film. The second strategy chosen was to cut a ring around the hole, allowing the obstruction to simply fall out. This was successful, though still a slow operation. Machining the ring sufficiently for the obstruction to fall out (shown in Figure 6) took approximately 5 hours of direct machining time. Including the set-up and monitoring time, several days were needed. Therefore, quicker alternatives are under investigation as this is not suitable for volume production.

3.2 Assembly Task 2 Trials

Two options were considered for the manipulation of the spheres, to allow it to be joined to the shaft: use of a microgripper to hold the sphere between two jaws and use of a single needle tip. Whilst the gripper is preferable for its positive hold on the sphere, it is also likely to create indentations, at the micro and nano-scale, to the sphere. This kind of damage is not acceptable as it will affect the performance of the completed micro-CMM probe. Therefore, it was concluded to try to work with a single needle tip to move the sphere. This technique required a great deal of operator skill and patience to achieve. A view of the assembly process is shown in Figure 7. The procedure followed to facilitate assembly is as follows: (1) Locate the sphere on the fixture and bring the needle tip to it. (2) Carefully push the sphere horizontally across the fixture surface to weaken the tension forces between the carbon and the sphere bond. (3) At a critical point, the tension forces between the sphere and needle exceed that of the sphere and the carbon and the sphere then becomes attached to the needle. (4) Manoeuvre the sphere to the end of the shaft for joining.

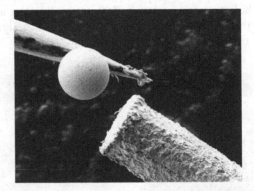

Fig. 7. SEM image showing the assembly of the sphere onto the shaft

This process does leave some residual adhesive on the surface of the sphere; however this was not visible in the FIB/SEM and should be easily removable with a non-abrasive cleaner. With the sphere in place, the next trials focused on joining the sphere to the shaft. The two main processes available were the use of a SEM-specific adhesive or the use of a FIB to deposit carbon tabs connecting the two parts. Since the adhesive would have to be manually applied to the shaft with the second manipulator, it was decided to test the feasibility of the FIB deposited tabs first. However, a number of factors adversely affect the potential success of such an approach. Firstly, FIB deposition is intended to be used at the sub-micrometre level and so producing tabs large enough to bridge the two parts in question would take in excess of twelve hours each. Secondly, as with all tab welds, it is necessary to hold both parts together and apply tabs from all sides – this is simply not possible within the vacuum chamber of the Zeiss NVision. Thus it became necessary to investigate the potential use of SEM glue. These adhesives are sufficiently viscous so as not to evaporate when placed in a vacuum chamber and are cured by a SEM beam. However, this also poses a significant challenge in giving limited time to complete the assembly as the SEM is used for viewing and guiding the process. As of yet the PMC has not achieved a sufficiently strong bond for the complete assembly to survive the rigours of the re-pressurisation process. However the feasibility of such an approach has been demonstrated.

4 Conclusion

This paper has presented the analysis and assembly planning and initial trials of an innovative micro-CMM probe. The work broke the assembly into two tasks, the assembly of the shaft to the flexure and the sphere to the shaft. Initials trials were begun for the former and multiple potential solutions were generated for the second. The results of these trials demonstrated that both solutions were feasible and worthy of more detailed examination in future research.

From the research conducted, several conclusions can be drawn:

- Utilising the Zeiss NVision for micro scale assembly of components is a feasible application, primarily due to the dexterity and accuracy of the micro manipulators.
- Microspheres can be manipulated and manoeuvred without the need for damaging grippers; however, this results in a substantial increase in the required effort, time, and operator skill to complete.
- Using FIB deposition to join two micro components is possible, but not feasible due to the time frames and the need for access to all sides of the assembly.
- Using SEM glue to join two micro components is feasible, but the process requires development and refinement.
- The FIB can be used to machine on the micro-scale, though the time frames are too great for volume production

4.1 Further Work

In order to facilitate the production of working prototype probes, the research efforts will focus on:

- A detailed examination of the performance of different adhesives and joining methods.
- Development of the use of the Kleindiek manipulators within the Zeiss NVision and investigation into their use outside of the chamber.
- Seeking alternatives to machining the flexure hole with the Zeiss NVision

Acknowledgements

The authors would like to acknowledge the inputs from P. Wentworth. The authors would also like to acknowledge the EPSRC 3D-Mintegration Grand Challenge Project, the National Measurement System Engineering Measurements Programme (2008-2011), the Centre of Excellence in Customised Assembly (CECA) HEFCE HEIF3 project and the East Midlands Development Agency funded Ultra Precision Machining Centre project, which provided the funding for this research.

References

1. Tietje, C., Ratchev, S.: Design for Microassembly – A Methodology for Product Design. In: International Symposium on Assembly and Manufacturing, pp. 1-4244-0563-7/07 (2007); Proceedings of the 2007 IEEE
2. Ratchev, S., Koelemeijer, S.: Micro-assembly technologies and applications. In: Fourth International Precision Assembly Seminar. International Federation for Information Processing, Chamonix, France (2008)
3. Peggs, G.N., Lewis, A.J., Leach, R.K.: Measuring the metrology gap - 3D metrology at the mesoscopic level. Journal of Manufacturing Processes 6(1), 117–124 (2004)
4. Stoyanov, S., Bailey, C., Leach, R., Hughes, B., Wilson, A., O'Neill, W., Dorey, R., Shaw, C., Underhill, D., Almond, H.: Modelling and Prototyping the Conceptual Design of 3D CMM Micro-probe. In: Proceedings of the IEEE 2nd Electronics System Integration Technology Conference, pp. 193–198 (2008)
5. Haitjema, H., Pril, W.o., Schellekens, P.H.J.: Development of a silicon-based nanoprobe system for three-dimensional measurements. CIRP Annals - Manufacturing Tech. 50, 365–368 (2001)
6. Van Brussel, H.P.: Assembly of Microsystems. CIRP Opening Session. Annals of the CIRP 49/2/2000 (2000)
7. Rampersad, H.: Integrated and Simultaneous Design for Robotic Assembly. John Wiley & Sons Ltd., Chichester (1994)
8. Smale, D., Ratchev, S., Segal, J., Leach, R.K., Claverly, J.D.: Assembly of the stem and tip of an innovative micro-CMM probe. In: Euspen 9th International Lamdamap (Laser Metrology and Machine Performance) Conference, Brunel, London, UK (2009)

Chapter 5

Metrology and Control for Micro-assembly

Miniaturized Camera Systems for Microfactories

Timo Prusi, Petri Rokka, and Reijo Tuokko

Tampere University of Technology, Department of Production Engineering,
Korkeakoulunkatu 6, 33720 Tampere, Finland
{timo.prusi,petri.rokka,reijo.tuokko}@tut.fi

Abstract. This paper presents our work on finding an alternative for standard machine vision equipment to be used in microfactories where small working spaces raise the need for miniaturized equipment. We tested three commercially available miniaturized camera modules used, for example, in mobile phones and compared them against two standard machine vision cameras. In the tests, we compared four selected factors: camera dynamic capability, image distortions, edge sharpness, and smoothness of image brightness.

Keywords: Microfactory, desktop factory, machine vision, miniaturized camera systems.

1 Introduction

Desktop and microfactory equipment refers to manufacturing and assembly equipment that can be placed on desktop and moved easily by human power. For example, the microfactory concept developed at Tampere University of Technology (TUT) [1] uses stand-alone factory modules that have dimensions of 300 x 220 x 200 mm and that have a working envelope of 180 x 180 x 180 mm. Because of the very small working envelope, all used equipment needs to be highly miniaturized. Fig. 1 shows the TUT microfactory module where the left part of the module is reserved for control electronics and the larger part on the right is the work envelope.

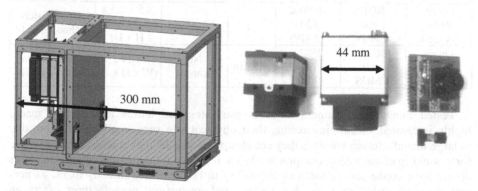

Fig. 1. TUT microfactory module (left) and tested C-mount cameras (without lenses), 2 megapixel, and 5 megapixel camera modules (right). Green circuit board under 5 megapixel module is an adapter board for evaluation purposes.

S. Ratchev (Ed.): IPAS 2010, IFIP AICT 315, pp. 115–122, 2010.

Typical assembly and manufacturing operations implemented in desktop and microfactories need or at least benefit greatly from the use of machine vision. Cameras can, for example, locate parts to be assembled, make dimensional measurements, or perform other quality assurance tasks. However, integrating standard machine vision cameras with standard C-mount or even S-mount optics to the small working envelope of a microfactory is extremely difficult because of their relatively large size. Therefore smaller cameras with smaller optics are needed. In this paper we present our work on finding an alternative for normal machine vision equipment.

2 Tested Cameras and Camera Modules

We tested three miniaturized camera modules used, for example, in mobile phones and compared them against two normal machine vision cameras with C-mount optics. Table 1 and Fig. 1 show tested cameras. In the tests, we used a C-mount lens with nominal focal length of 12 mm (type JHF12MK) from SpaceCom [2]. The length of the C-mount lens (about 36 mm for the lens used) is not included in the physical size mentioned in Table 1 but it has to added to depth length of C-mount cameras. For camera modules, the physical depth dimension includes the integrated lens.

Table 1. Tested C-mount cameras and camera modules

Camera	Type	Resolution	Pixel Size	Full Well Capacity	Physical Size (mm)	Manufacturer
UI-1540-M (C-mount)	Grayscale (CMOS, USB)	1280 x 1024 (SXGA, 1.3 MP)	5.2 µm	40 000 e-	32 x 34 x 38 (W x H x D)	Imaging Development Systems [3]
UI-6240-SE-M (C-mount)	Grayscale (CCD, GigE)	1280 x 1024	4.65 µm	12 000 e-	44 x 34 x 60 (W x H x D)	Imaging Development Systems [3]
Omni-Vision OV-07640	RGB color, CMOS	640 x 480 (VGA, 0.3 MP)	4.2 µm	35 000 e-	6 x 6 x 5 (W x H x D)	OmniVision Technologies [4]
Omni-Vision OV-2640	RGB color, CMOS	1600 x 1200 (2 MP)	2.2 µm	12 000 e-	8.5 x 8.5 x 5.5 (W x H x D)	OmniVision Technologies [4]
Omni-Vision OV-5620	RGB color, CMOS	2592 x 1944 (5 MP)	2.2 µm	Not known	21 x 19 x 16 (W x H x D)	OmniVision Technologies [4]

Tested miniaturized camera modules use integrated lenses and custom made, highly integrated, electronics making their physical size very small. In addition, due to large manufacturing volumes they are cheap making them an interesting alternative for normal machine vision equipment. As such, tested miniaturized camera modules do not have connectors or software capability to be connected directly to PC as normal machine vision cameras. For testing and evaluation, manufacturer offers an evaluation kit with USB and/or Ethernet connectors and software enabling connection to PC.

3 Test Targets and Image Capturing

We tested each camera to find out 1) how much geometrical distortions images have, 2) how uniform image brightness is, 3) how sharp edges images have, and 4) how well cameras can detect dark and bright objects at the same time (dynamic capability). For this purpose, we used targets shown in Fig. 2. Test targets were: a) a checker board pattern for calibrating and calculating image geometrical distortions, b) uniform mid gray (pixel value 128, max 255) for checking brightness uniformity, c) two patterns with black and white bars and slanted squares for evaluating edge sharpness, and d) a pattern with 17 regularly distributed grayscales ranging from completely black (pixel value 0) to completely white (pixel value 255) for evaluating the dynamic capability of the imaging system. Final test target having some common objects is only used for visual estimations.

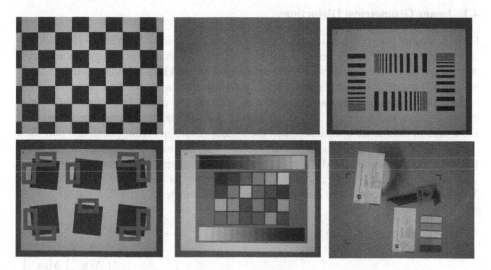

Fig. 2. Test targets imaged with OmniVision VGA camera module. Red rectangles on top of slanted square target indicate areas where edge sharpness was evaluated.

The targets were printed on normal A3 and A4 size papers with high quality color laser printer using 1200 dpi printing resolution. We took three images of each target in a room with no windows and normal office illumination created with fluorescent tubes in the ceiling. When taking images, we adjusted the distance between camera and target so that the field-of-view (FOV) was always slightly over 300 mm wide fitting A4 size paper. We also took care that the targets were always in the same orientation. With C-mount cameras we used the same lens with same aperture size. Before and after taking images, we measured illumination intensity in FOV corners and center with an exposure meter commonly used in photography. To avoid effects created by image compression, we saved all images in bitmap format. Table 2 summarizes imaging conditions.

Table 2. Imaging conditions when taking test images

	UI-1540	UI-6240	Omni VGA	Omni 2 MP	Omni 5 MP
Lens-target distance (mm)	584	656	393	371	487
Camera integration time (ms)	119	57	Not known (auto-exposure)	Not known (auto-exposure)	Not known (auto-exposure)
Illumination intensity (exposure values)	7.5 – 7.7	7.5 – 7.7	7.4 – 7.6	7.4 – 7.6	7.4 – 7.6

4   Analysis and Test Results

4.1   Image Geometrical Distortions

For calculating image distortions, we used calibration toolbox for Matlab [5] to detect the corners of the checker board pattern to get a 9 x 6 matrix of image coordinates. After that, we used calibration method developed by Heikkilä [6] and implemented in Matlab [7] to calibrate the camera + lens system and to calculate the corrected corner image coordinates. Finally, we calculated the distance in pixels between the original, measured, and the corrected image coordinates. Knowing the camera resolution and FOV size in millimeters, we calculated the spatial resolution (mm per pixel) and used that to transform pixel distances to millimeters. Fig. 3 shows these errors in graphical format for OmniVision 5 MP camera module. The shape of the error pattern was similar for all cameras and camera modules: largest errors are in corners. Table in Fig. 3 shows the average and maximum distortions in millimeters.

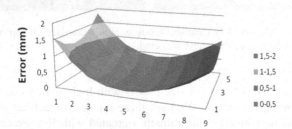

Geometrical distortions, Omni 5 MP

	Avg (mm)	Max (mm)
UI-1540	0.47	1.42
UI-6240	0.32	1.19
OV VGA	0.32	1.37
OV 2 MP	0.74	1.51
OV 5 MP	0.51	1.82

Calibration grid corner points

Fig. 3. Geometrical distortions in 9 x 6 calibration points for OmniVision 5 MP camera module and average and maximum distortions for all tested cameras and camera modules

4.2   Image Brightness Uniformity

To evaluate the uniformity of image brightness, we divided images to 10 x 8 equally sized windows and calculated the average pixel intensities for each window. In these calculations, we did not consider the small variations in illumination intensities in different parts of camera FOV, because they were constant and small. We scanned the target with a normal desktop scanner using 600 dpi resolution and made the same analysis for the scanned image to verify target brightness. Even though the test target was printed with high quality laser printer to constant mid gray color (pixel value 128, max 255), it proved to have small variations in measured gray values (brightnesses). As leftmost graph in Fig. 4 shows, scanned target brightness changed relatively randomly whereas OmniVision 2 MP images were substantially brighter in center part of the image.

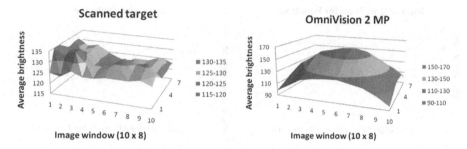

Fig. 4. Measured average pixel brightnesses for 10 x 8 image windows of scanned target (left) and for OmniVision 2 MP image (right). Graphs have different scales in Z direction.

Table 3 lists minimum, average, and maximum brightness values for scanned target and for all cameras and camera modules. Images captured with C-mount cameras had relatively even brightness whereas all camera modules produced images where the center part of the image was noticeably brighter than corners and edges.

Table 3. Image brightness uniformity test results for all cameras and camera modules

	Scanned	UI-1540	UI-6240	OV VGA	OV 2 MP	OV 5 MP
Min	121.0	96.0	96.0	71.0	96.0	74.3
Average	127.0	104.3	101.2	98.8	132.8	110.3
Max	135.0	113.0	115.0	118.7	161.0	134.7
Standard deviation	3.4	4.2	4.9	12.1	15.9	16.9
Max - Min	14.0	17.0	19.0	47.7	65	60.3

4.3 Edge Sharpness

We used ImaTest software [8] to calculate two measures for edge sharpness from 10 different positions from the slanted squares test target (see Fig. 2). First measure is Modulation Transfer Function (MTF) and especially MTF50 value. MTF50 value refers to frequency when contrast between input and output has dropped to 50% of its original value. In practice this means, for example, distance between black and white bars where contrast between black and white has dropped to 50% of original making bars seem blurry. ImaTest's SFR function calculates MTF50 value in line widths and divides it by picture height to compensate different picture resolutions. Second measure for edge sharpness is the distance, measured in pixels, from background to target pixel brightness values giving the "steepness" of the edge. ImaTest calculates 10% - 90% rise distance and scales it to picture height as in MTF50 calculations.

Fig. 5 shows MTF50 value in line widths (LW) divided by picture height and 10% - 90% edge rise distance also scaled with picture height (PH). In both graphs, the higher the value the better it is. Horizontal axis in both graphs refers to the 10 different edges where, for example, LTV means Left Top corner Vertical edge and MiH means Middle square and Horizontal edge (see Fig. 2).

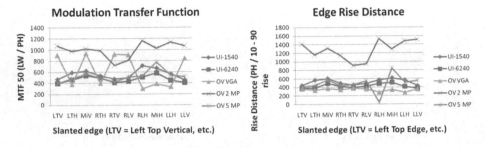

Fig. 5. Edge sharpness results

Tested camera modules use Bayer mosaic filter to detect colors. This commonly used technique is simple but, as a drawback, it adds (colored) artifacts around target edges. Tsai and Song [9] explain this in detail and propose a method to reduce color artifacts. Tested camera modules clearly do not use such methods as all images captured with all camera modules show these artifacts as colors next to target edges on black and white targets. Fig. 6 shows a close-up on bar pattern imaged with OV 2 MP module. Graph on right in Fig. 6 shows red, green, and blue pixel intensities measured in horizontal direction. This graph shows that red, green, and blue pixel values have peaks in slightly different positions making some pixel columns seem colored.

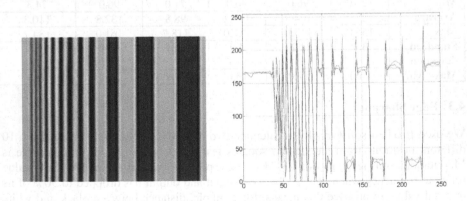

Fig. 6. Close-up on edge pattern showing colored artifacts

4.4 Dynamic Capability

We evaluated camera dynamic capability by using a pattern with 17 different gray-scales ranging from completely white (pixel value 255) to completely black (pixel value 0) with even steps. ImaTest's Stepchart feature calculated average pixel values for each 17 steps for all cameras, camera modules, and also for scanned target. Fig. 7 shows these results. When taking images with C-mount cameras, we adjusted camera integration time so that the white step would be almost overexposed (pixel value almost 255). Camera modules used automatic exposure and therefore white steps do not appear completely white but have pixel values around 180.

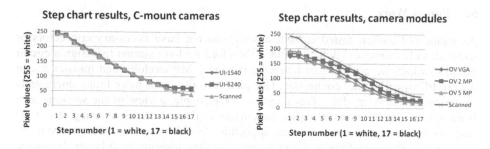

Fig. 7. Measured image brightnesses of 17 gray steps for C-mount cameras (on left) and for camera modules (on right)

5 Discussion and Conclusions

Based on these tests, tested miniaturized camera modules provide good enough quality images to be, in most applications, realistically comparable with standard machine vision equipment. Even though tested camera modules have higher pixel resolutions than C-mount cameras, one has to remember that using Bayer mosaic filter reduces "true image resolution" because pixel (color) values are interpolated from several neighboring pixels. Another drawback resulting from Bayer filter are the (colored) artifacts on target edges possibly making edge detection more difficult than when using monochrome (grayscale) cameras. Edge sharpness and image geometrical distortions were similar in all tested cameras and camera modules. Variations in image brightness are probably the most significant difference between tested cameras and camera modules.

One fundamental difference between standard machine vision cameras and miniaturized camera modules is the level of automatization: Camera modules have several automatic software features manipulating raw camera image before outputting it; for example auto exposure setting, automatic white balance, and edge enhancement are just a few features camera modules automatically adjust. Such features are convenient when the goal is to (automatically) make images look good to human eye. In typical machine vision applications, however, we want to control, or at least know, what parameters were used when image was captured in order to able to reliably compare images from the same scene. Tested camera modules have limited and/or poorly documented methods to control image capturing parameters and, considering their use in machine vision applications, this is a definite weakness for them. Other weaknesses are their short lifespan and limited availability and support at least for small customers.

On the other hand, considering desktop and microfactory applications, the extremely small size of miniaturized camera modules is a distinctive advantage. Small size enables easy integration and placing cameras to places where normal machine vision cameras are impossible to fit. Further advantage of camera modules is their low price: the modules tested here cost approximately 20 € per piece. Therefore it would be economically feasible to use multiple cameras in each microfactory module enabling completely new ways of monitoring and measuring production.

5.1 Future Work

As mentioned earlier, tested camera modules do not have necessary connectors or software to be connected directly to PC. Therefore we have started to design a circuit board to which we can connect four OmniVision 2 MP modules and transfer image data to PC over Ethernet connection. Our plan is to fit one or more four camera units in our microfactory module. This gives us, for example, a view of the working area from several directions enabling measurements in three dimensions using stereo vision and/or photogrammetry. Second application could be to use different exposure settings in cameras looking at the same area enabling imaging with better dynamics, i.e. detecting very bright and dark objects at the same time. Third possibility is to combine several partially overlapping images into one high resolution image. Researchers at Stanford have implemented these using up to 128 conventionally sized cameras [10]. Our aim is to achieve similar results in microfactory environment using miniaturized camera modules.

References

1. Heikkilä, R., Karjalainen, I., Uusitalo, J., Vuola, A., Tuokko, R.: The Concept and First Applications of the TUT-Microfactory. In: 3rd International Workshop on Microfactory Technologies, Seogwipo KAL Hotel, Jeju-do, Korea, pp. 57–61 (2007)
2. Space Inc lenses, http://www.spacecom.co.jp/english/
3. IDS Imaging Development Systems cameras, http://www.ids-imaging.com/
4. OmniVision miniaturized camera modules, http://www.ovt.com/
5. Bouguet, J-Y.: Camera Calibration Toolbox for Matlab, http://www.vision.caltech.edu/bouguetj/calib_doc/
6. Heikkilä, J.: Geometric Camera Calibration Using Circular Control Points. IEEE Transactions on Pattern Analysis and Machine Intelligence 22(10), 1066–1077 (2000)
7. Heikkilä, J.: Camera Calibration Toolbox for Matlab, http://www.ee.oulu.fi/~jth/calibr/
8. Imatest software for testing digital image quality, http://www.imatest.com/
9. Tsai, C.-Y., Song, K.-T.: A new edge-adaptive demosaicing algorithm for color filter arrays. Image and Vision Computing 25(9), 1495–1508 (2007)
10. Stanford Multi-Camera Array, http://graphics.stanford.edu/projects/array/

Vision and Force Sensing to Decrease Assembly Uncertainty

R. John Ellwood, Annika Raatz, and Jürgen Hesselbach

Institute for Machine Tools and Production Technology, TU-Braunschweig
Langer Kamp 19b, 38106 Braunschweig, Germany
j.ellwood@tu-bs.de

Abstract. This paper presents two ways of decreasing the assembly uncertainty of micro assembly tasks through further or optimized integration of sensors within a size adapted assembly system. To accomplish this, the orientation of the part to be placed with respect to the vision sensor is changed. This was possible through a new gripper which was able to overcome the restrictions placed on the system by the vision sensor. Another increase in precision was obtained through the integration of a force sensor into the wrist of the robot. This force sensor provides additional information about the placing process which allows the maximal force in the vertical axis to be limited. These improvements are then demonstrated on a task which requires the placement of linear guides which are 8.4 millimeters by 1 millimeter.

Keywords: Precision assembly, sensor guidance, size adapted robot.

1 Introduction

Micro assembly deals with the joining of parts which have at least one dimension less than 1 millimeter which need to be placed with accuracy on the order of a couple of micrometers or less. To accomplish such tasks, the hybrid parallel robot micabo^{f2} was designed and constructed at the Institute for Machine Tools and Production Technology at the TU- Braunschweig [1]. The main task of this robot has been the assembly of a linear micro actuator. Although capable of precision assembly, the goal of improving the abilities of the micabo^{f2} is an active area of research.

Within this paper the precision assembly robot micabo^{f2} is presented. Machine vision is accomplished through a 3D vision sensor which has been integrated into the head unit of the micabo^{f2}. This is followed by a brief overview of the precision assembly task which is used as a demonstration assembly task within the scope of this paper. With an understanding of the current system, two short coming are addressed, than overcome through the integration of additional sensors. It is ultimately shown that the precision as well as robustness of the micabo^{f2} for assembly tasks is improved through these sensors.

The first method strives to take advantage of a better orientation of the rectangular part being assembled within the rectangular field of vision of the 3D vision sensor. It was shown in [2] that the limited view of the part during relative positioning increases the assembly uncertainty.

S. Ratchev (Ed.): IPAS 2010, IFIP AICT 315, pp. 123–130, 2010.

The second method looks at limiting the forces that occur at the end of the gripper when a part is being placed. These irregular forces can arise from things such as differences associated with part tolerances or different aspects of the glue being used. Under these loads, small parts can easily slide or deform. To overcome these effects, a force sensor is integrated into the wrist of the micabo^{f2}. Its integration into the robot, along with the controller which is realized is presented.

1.1 Robot

The micabo^{f2} is a size adapted parallel robot which uses two linear actuators to move two arms which meet to form the robot head figure 1. This planar configuration allows the robot 2 degrees of freedom (dof), and is the basis for it's the high precision and accuracy. Located within the head unit of the robot are 3 more dof which allow the end effector to rotate, the vertical actuation of the gripper, as well as an actuator to focus the vision sensor. The workspace of the robot is 160mm x 400mm x 15mm. Accuracy measurements which were performed according to EN ISO 9283 [3], show that the micabo^{f2} achieved a repeatability of 0.6 micrometer.

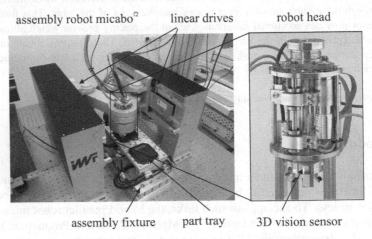

Fig. 1. The size adapted robot, micabo^{f2}

1.2 Three Dimensional Vision Sensor

The 3D vision sensor implemented within the micabo^{f2} was developed at a partner institute at the Technical University Braunschweig, the Institute of Production Measurement Engineering (IPROM) [4]. As can be seen in figure 2, the sensor takes advantage of stereo photogrammetry to gather 3D data. Here the stereo view of the environment requires only one camera, which allows the entire sensor to be fitted into the head of the robot. The sensor has a field of vision 11mm by 5.5 millimeters and offers a resolution of 19 micrometers/pixel. Repeatability measurements have shown that this sensor has a standard deviation of σ_x = 0.220 micrometer and σ_y = 0.290 micrometer.

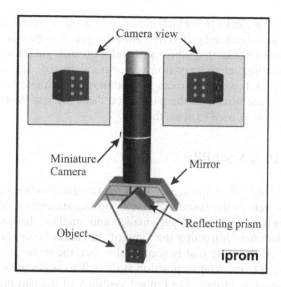

Fig. 2. Functional principle of the 3D vision sensor

1.3 Precision Assembly of Active Microsystems

The goal of the Collaborative Research Center 516 is the design and construction of a micro linear stepping motor [5]. The challenges in developing and manufacturing this reluctance based motor are equaled by the task of assembling it with the required tolerances. To overcome these, the mentioned size adapted robots along with its 3d vision sensor are being optimized to best fulfill this task. The task of improving the placement of these guides on the surface of the stator element is one of the main motivations for the presented work and is thus used as a demonstration task. The simplified motor model for this assembly task can be seen in figure 3, along with dimensions.

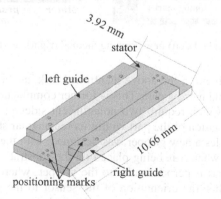

Fig. 3. Simplified linear stepping motor assembly

Circular positioning marks [6], which are created using a photolithographic manufacturing process, have been integrated into both the guides to be placed as well as the stator. The positioning marks (8 on the stator and 4 on each guide) in both images are measured and a resulting relative position vector is calculated. This can then be used within the robot control, which corrects the rotational and translational error within the plane parallel to the surface of the part. Once these are less than 0.8 micrometer, the vertical information is used to place the part.

2   Increasing Part Visibility

A limiting factor of the 3D vision sensor used for relative positioning is that the part being handled, as well as the functional marks on the substrate always needs to be visible. As the parts to be handled get smaller and smaller, the task of effectively gripping them, while not obstructing the view of the marks becomes more and more challenging. Further, the more that is seen of the part, the more accurately it can be placed using a vision based relative position system. This was shown in [2], in which a large assembly error is attributed to limited visibility of the part during the relative positioning phase. It is shown that the assembly error of the finished part was drastically larger on the side of the part which was covered by the gripper.

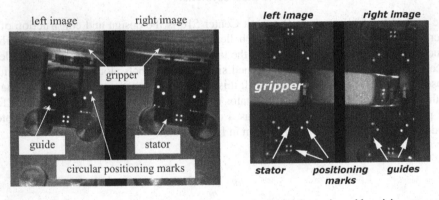

Fig. 4. Old gripping method (left) new gripping method (right), as viewed by vision sensor

Herein lays the problem in the above mentioned task, as there are limited ways of gripping such a long and narrow part. This is further complicated by the limitations of the 3D vision sensor which require two nonadjacent sides of the part be visible, as well as limits the orientation of the part within its rectangular shaped sensing area. To overcome these obstacles a new gripper was designed, with the goal of improving the visibility of the stator when it is being placed on the substrate. This new gripper was designed so that the part is perpendicular to the gripper, where the old gripper had it parallel. Along with this, the orientation of the sensor to the gripper was rotated 90 degrees, so that the entire stator as well as guide can be seen during the placing process. As can be seen in figure 4, the combination of the new gripper and orientation of the part under the sensor allow the entire part to be seen during the place phase of the assembly task.

To facilitate a quantitative comparison between the old and new gripper, parts were assembled using the new gripper and compared against those with the old gripper. The resulting assembly data was then used to calculate the positioning uncertainty and assembly uncertainty. According to DIN ISO 230-2 [7], the positioning uncertainty represents the relative position error between the assembly parts before the bonding process, while the assembly uncertainty is measured after the bonding has been completed. Both terms are calculated as a combination of the mean positioning deviation and the double standard deviation.

To recall the assembly data which was obtained through the assembly of 33 guide/stator groups and the older vacuum gripper, a positioning uncertainty of 1.2 micrometers and an assembly uncertainty of 36 micrometers were thus obtained [2]. The resulting assembly uncertainty along with the data can be seen on the left hand side of figure 5.

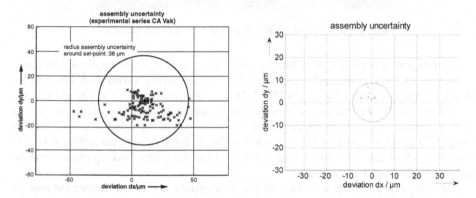

Fig. 5. Assembly uncertainty for the old gripper (left) along side the assembly uncertainty of the new gripper (right)

Using the newly developed gripper in conjunction with the new orientation, 10 guides were assembled on stators. This data showed a similar positioning uncertainty of 1.3 micrometers, while the assembly uncertainty was reduced to 8.6 micrometers. This data along with a circle representing the uncertainty can be seen on the right hand side of figure 5. Although this limited data does not allow for a conclusive statement, it does allow one to draw the inference that the new orientation decreases the assembly uncertainty.

3   Force Sensor Integration

Although information about the relative difference in height between the object being placed and the substrate on which it is being placed can be gained from the 3D vision sensor, it is unable to overcome differences which arise from inconsistent glue drops as well as part tolerances. In order to gain more information in the vertical axis, a force sensor was added into the wrist of the micabo^{f2} robot.

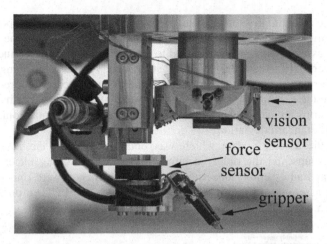

Fig. 6. Force sensor integrated into the wrist of the micabo[f2]

An integral part of this was finding the correct sensor, as it required a balance between size, force range, and sensitivity. These characteristics were found best satisfied with the Schunk FT Nano-1712/0.12. With a maximal force of ±17N and resolution of ±1/160 N in the Z axis, it was viewed as adequate for this application. It was then integrated between the gripper and the vertical axis of the robot as can be seen in figure 6. Though this sensor has a coarse resolution for precision assembly, it is fine enough to allow the robot to gain information about the contact taking place between the part being placed and substrate.

After the sensor's integration into the physical system, a software calibration routine was developed which takes the offset of the sensor to the gripper into consideration. This was achieved through a force and moment balance, ultimately allowing force on the gripper to be solved for. The last step was the integration into the control loop. Here a cascade control loop is chosen with the inner loop controlling position, and the outer adjusting the position as a function of the force [8,9]. The resulting control loop can be seen in figure 7. A saturation function was introduced which prevents the robot from moving down to "increase" the force if it is less than the desired force.

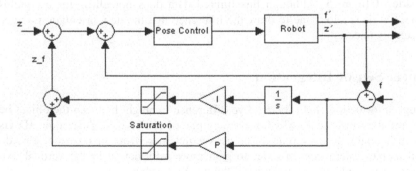

Fig. 7. The implemented cascade force/position controller

After the functionality of the force controller was confirmed, it was used to set square parts on the above mentioned stator in conjunction with a hot melt adhesive. With its implementation, it is shown that the robot system is better able to accommodate irregularities which can arise when hot melt adhesives are used. Here the properties of the hot melt adhesive during the placing portion, such as whether or not the drops are fully melted or not can be determined. In the case that a larger droplet of the hot melt is not thoroughly heated and thus still hard, the force controller can limit the maximal force seen at the gripper. Reducing the maximum seen force at the gripper not only prevents damage to the robot or part but also improves the precision, as parts are less likely to move while being placed.

4 Conclusions

Two ways of improving the precision of a micro assembly using a size adapted parallel robot were introduced. First the shortcomings of the 3D vision sensor were overcome through the integration of a new gripper in conjunction with rotating the sensor to take advantage of its rectangular image sensor. It was shown that this resulted in an assembly uncertainty of 8.6 micrometers, in comparison with the old uncertainty of 36 micrometers. Next a brief review of how a force sensor was integrated into the Micabof2 was presented. This sensor was used to provide redundant information about the vertical axis, ultimately improving the robustness of the robot to accurately indifferences in the z axis.

Acknowledgements

The authors gratefully acknowledge the funding of the reported work by the German Research Center (Collaborative Research Center 516).

References

1. Simnofske, M., Schöttler, K., Hesselbach, J.: micaboF2 - Robot for Micro Assembly. Production Engineering XII(2), 215–218 (2005)
2. Schöttler, K., Raatz, A., Hesselbach, J.: Ratchev, S., Koelemeijerm, S. (eds.) Micro-Assembly Technologies and Applications. IFIP International Federation for Information Processing, vol. 260, pp. 199–206. Springer, Boston
3. EN ISO 9283, Industrieroboter, Leistungskenngrößen und zugehörige Prüfmethoden. Beuth-Verlag, Berlin (1999)
4. Tutsch, R., Berndt, M.: Optischer 3D-Sensor zur räumlichen Positionsbestimmung bei der Mikromontage. Applied Machine Vision, VDI-Bericht Nr. 1800, Stuttgart, pp. 111–118 (2003)
5. Hahn, M., Gehrking, R., Ponick, B., Gatzen, H.H.: Design Improvements for a Linear Hybrid Step Micro-Actuator. Microsystem Technologies 12(7), 646–649 (2006)
6. Berndt, M., Tutsch, R.: Enhancement of image contrast by fluorescence in microtechnology. In: Proceedings of SPIE, Optical Measurement Systems for Industrial Inspection IV, München, vol. 5856, pp. 914–921 (2005)

130   R.J. Ellwood, A. Raatz, and J. Hesselbach

7. DIN ISO 230-2, Prüfregeln für Werkzeugmaschinen, Teil 2: Bestimmung der Positionierunsicherheit und der Wiederholpräzision der Positionierung von numerisch gesteuerten Achsen. Beutch Verlag, Berlin (2000)
8. Caccavale, F., Natale, C., Siciliano, B., Villani, L.: Integration for the next generation, embedding force control into industrial robots. In: Proc. IEEE International Conference on Robots and Automation, Orland Florida (May 2006)
9. Chiaverini, S., Siciliano, B., Villani, L.: Force and position tracking: Parallel control with stiffness adaption. IEEE Control Systems, 27–33 (February 1998)

Modelling the Interaction Forces between an Ideal Measurement Surface and the Stylus Tip of a Novel Vibrating Micro-scale CMM Probe

J.D. Claverley[1], A. Georgi[2], and R.K. Leach[1]

[1] National Physical Laboratory, Hampton Road, Teddington, UK, TW11 0LW
Tel.: +44 20 8943 6242
james.claverley@npl.co.uk
[2] The University of North Carolina at Charlotte, NC, USA

Abstract. This paper describes the development of an analytical model to describe a novel three-axis vibrating micro-scale probe for micro-co-ordinate measuring machines (micro-CMMs). The micro-CMM probe is vibrated in three axes, in order to address the problems inherent with micro- and nano-scale co-ordinate measurements caused by surface interaction forces. These surface forces have been investigated and a mathematical model describing an ideal probing situation has been developed. The vibration amplitude required for the probe to overcome the surface interaction forces has been calculated using this model. The results of initial vibration experiments are reported and the suitability of the probe to counteract the surface interaction forces is confirmed.

Keywords: micro-CMM probe, surface interaction forces.

1 Introduction

As micro- and nano-scale engineering becomes ever more prevalent in many aspects of modern manufacturing, there is an urgent need for accurate and traceable micro- and nano-metrology. Recent advances in co-ordinate metrology, specifically the development of micro-CMMs, have enabled significant improvements to be made in three dimensional traceable micro- and nano-metrology. These micro-CMMs have highly advanced and rigid metrological frames and high accuracy measurement scales capable of nanometre resolution. The measuring accuracy of micro-CMMs has been investigated and found to be limited by the probe systems, rather than the machines themselves [1]. The National Physical Laboratory (NPL) is continuing to work on advanced micro-CMM probe design with the aim of improving the overall accuracy of modern micro-CMMs to less than 100 nm [2].

Simply reducing the size of current CMM probes is no longer a viable method of producing an accurate micro-CMM probe. Such reduction in size has been taken to its limit by various research projects both within NPL [2] and at other laboratories around the world [3]. Any work done to continue the advancement of micro-CMM probe research should implement a total redesign to take into account the need for low contact probing forces and the effect of the surface interaction forces. Research that has addressed some of these issues includes the development of a high aspect ratio

S. Ratchev (Ed.): IPAS 2010, IFIP AICT 315, pp. 131–138, 2010.

fibre-based probe for 2D co-ordinate metrology, low force silicon based microprobes with piezoresistive strain gauges, CMM probes that optically measure probe deflection and ultra low force scanning probes that use laser trapped micro spheres (see reference [3] for a review of current probing technology).

To address the effects of surface forces and the need for low force probing, a vibrating micro-CMM probe has been designed and produced. It consists of a triskelion (three legged) flexure device and a micro-stylus [4]. The vibration of the probe will be controlled so that the stylus tip is always vibrating normal to the measurement surface and so that the acceleration of the stylus tip is sufficient to counteract the surface interaction forces. The vibrating aspect of this probe enables it to work in a non-contact fashion that will, theoretically, reduce the probing force incident on the measurement surface to zero.

The micro-CMM probe is made to vibrate by using six piezoelectric (PZT) thin film actuators (two on each flexure) and two PZT sensors on either end of each flexure which detect any change in vibration amplitude. The size of the flexures, the size, shape, and sensitivity of the sensors, and the ability of the triskelion device to exert forces isotropically on the measurement surface, have been subject to extensive research. The basic design of the new vibrating micro-CMM probe is shown in Fig. 1.

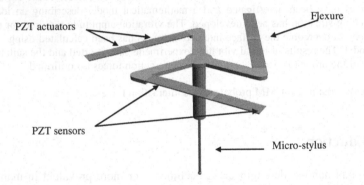

Fig. 1. A schematic diagram of the vibrating micro-CMM probe

2 Surface Force Theory

An analytical model was created to describe the effects that the surface forces will have on the oscillatory behaviour of the probe and to determine whether the amplitude of the probe was sufficient to counteract the surface interaction forces.

It is assumed that the device can be directly compared to a driven oscillator subject to a damping force. The second order differential equation that describes this situation is

$$m\frac{\mathrm{d}^2 z}{\mathrm{d}t^2} + b\frac{\mathrm{d}z}{\mathrm{d}t} + kz = F_0 \cos(\omega t) \qquad (1)$$

where m is the mass of the oscillator, z is the position of the bottom of the probe tip, b is the damping coefficient, k is the spring constant of the oscillator, F_0 is the driving force and ω is the angular frequency. The spring constant of the probe has been found to be 15 N·m^{-1} using a finite element model.

A good understanding of the oscillatory behaviour of the probe can be gained from equation 1. To solve for z, an appreciation of b, and hence the surface forces, is required. A set of models was used to represent the surface interaction forces and a more complete model of the changing frequency and amplitude of the probe tip was calculated using equation 1. Note that a similar investigation has been conducted for a high accuracy vibrating probe being developed by the University of North Carolina at Charlotte [5].

2.1 The Capillary Force

Capillary forces play a dominant role in micro- and nano-scale interactions [6]. The basis of the capillary force is the thin liquid film layers that accumulate on surfaces from condensation and contamination. As two surfaces are brought together the films join and the surface tension causes an adhesion effect. At the micro- and nano-scale these forces can often be the dominant force governing interactions. Contact angle plays a large role in determining the magnitude of capillary forces and can be affected by the material surface chemistry as well as the surface form and texture. A schematic representation of contact between a sphere and a surface with a liquid layer is shown in Fig. 2.

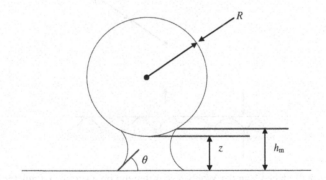

Fig. 2. A schematic representation of a sphere, radius R, interacting with a surface with a liquid layer. The diagram shows the variables z (*distance from surface*), h_m (*height of meniscus*) and θ (*the contact angle*).

The equation used to describe the capillary force between a sphere and a plane surface is

$$F_1 = \frac{4\pi R \gamma_\mathrm{L} \cos \theta}{1 + \dfrac{z}{h_\mathrm{m} - z}} \tag{2}$$

where γ_L is the surface tension of the liquid [6].

2.2 The Electrostatic Force

Electrostatic forces result from electrical charge generation or charge transfer between two surfaces. Under most circumstances the amount of charge separated is small due to the breakdown strength of the surrounding air. However, as gap size decreases below the mean free path for air (approximately 1 μm), the magnitude of the charge density can increase by orders of magnitude [7]. The strength of the electrostatic force also depends strongly on the material characteristics of the interacting bodies. The electrostatic force is determined by the electrical potential difference of the two interacting surfaces and is given by

$$F_2 = \frac{\varepsilon_0 U^2 \pi R^2}{z^2} \tag{3}$$

where ε_0 is the permittivity of free space and U is the potential difference between the two interacting surfaces [8].

2.3 The Van der Waals Force

Van der Waals forces arise from the polarisation of atoms and molecules as they are drawn together. A schematic representation of contact between a sphere and a surface is shown in Fig. 3.

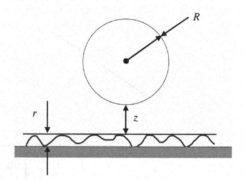

Fig. 3. Reduced Van der Waals forces depend on a surface parameter r

To determine the Van der Waals force an appreciation of the Hamaker constant [9] for the interacting surfaces must be gained. There are multiple methods of calculating the Hamaker constant all of which require empirical data. Additionally, Van der Waals forces can be significantly altered by surface roughness. An approximation of the Van der Waals force that includes the effects of surface roughness is given by

$$F_3 = \left(\frac{2z}{2z + r} \right)^2 \frac{HR}{6z^2} \tag{4}$$

where r is a constant describing the average surface roughness of the two surfaces and H is the Hamaker constant for the materials involved [10].

2.4 The Casimir Force

A prediction of quantum electrodynamics is that the exclusion of electromagnetic modes will cause two closely spaced conducting plates to be mutually attracted at the micro-scale [11]. Even though this attraction is extremely weak, especially between a sphere and a plate (as opposed to two parallel plates), when the separation distance is below 1 μm, the Casimir force tends to be stronger than the Van der Waals force and the electrostatic force. The equation to describe the Casimir force between a sphere and a plane is

$$F_4 = \frac{R\pi^3 hc}{360z^3} \tag{5}$$

where h is Planck's constant divided by 2π and c is the speed of light in a vacuum.

2.5 The Squeeze-Film Effect

Any micro device not enclosed within a vacuum will experience damping due to the finite viscosity of air [12]. This effect is non-linear and is assumed to be negligible in the current model, but its effects will be included in any future models.

3 Surface Interaction Model

The micro-CMM probe was designed such that its amplitude actively counteracts the influence of any forces it is subject to while interacting with a measurement surface. These surface forces, as described in the previous section, are dependant on certain physical attributes of the system. These attributes include (but are not limited to) the densities of the two interacting materials (the probe material and the measurement surface material), the Hamaker constant for the two interacting materials and air, the depth of the combined surface imperfections of the two interacting materials, and the contact angle of any liquid contamination between the two interacting surfaces.

To simplify the model it was assumed that the materials used for interaction experiments would be similar – tungsten carbide in this case. Table 1 shows the assumed values of the physical constants used to populate the surface force and vibration model.

Table 1. Physical attributes of the probe system used in the probe-surface interaction model

Physical Attribute	Value
Density of tungsten carbide	15 000 kg·m^{-3} [13]
Radius of probe tip (R)	50 μm
Combined depth of surface imperfections (r)	100 nm
Contact angle of water with tungsten carbide (θ)	85°
Surface tension of water (γ)	0.072 75 J·m^{-2} [13]
Hamaker constant for tungsten carbide and air (H)	21 × 10^{-21} J [9]
Natural frequency of probe	1.6 kHz

The contact angle of water with tungsten carbide was measured using a Krüss DSA100M picolitre dosing system. The reported angle is an average value over several tungsten carbide gauge block surface samples. The depth of the surface imperfections is estimated as a worst-case value for a grade K tungsten carbide gauge block surface.

The analytical model has been constructed accounting for the four main surface interaction forces. In order to model the worst-case scenario, gravity was also included although it is generally regarded as insignificant at this scale. To approximate the total force that a probe would experience, the arithmetic sum of the individual surface forces was calculated. This force approximates the worst-case scenario in which all of the forces are acting against the probe and it is vibrating in the vertical direction. The minimum required amplitude of oscillation for the probe based on surface conditions, probe mass and oscillation frequency is

$$A_{\min} = \frac{2F_s}{k} \tag{6}$$

where F_s is the sum of all the constituent surface forces. The strength of the individual forces, F_1, F_2, F_3 and F_4, and the total surface interaction force, F_s, are shown in Fig. 4.

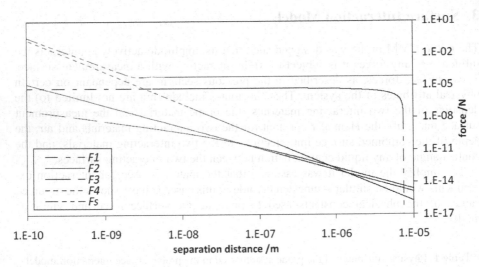

Fig. 4. Theoretical surface force strength with respect to separation distance, z

It can be seen that F_1 (the capillary force) dominates over a large range of separation distances. F_s, the combined surface force, includes the gravitational force on the finite mass of the device and is therefore an order of magnitude larger than F_1 at large separation distances.

By combining equations 2, 3, 4 and 5 to evaluate the total surface interaction force with respect to z (the separation distance between the probe and the surface), the minimum amplitude required to counteract the surface forces when probing 2 nm from the surface was found from equation 6 to be approximately 700 nm.

4   Results

Initial work on a micro-CMM probe remote from any surface investigated the amplitude of the vibration with respect to varying actuation voltages and offsets. The amplitudes were measured using a laser Doppler vibrometer. These results are shown in Fig. 5.

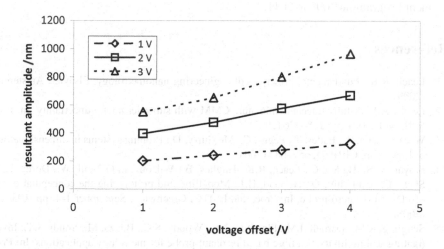

Fig. 5. Measured micro-CMM probe vibration amplitude with respect to voltage offset on the actuators for three different actuation voltage amplitudes

It can be seen from Fig. 5 that a wide range of amplitudes ranging from 200 nm to 1 µm can be achieved with the vibrating micro-CMM probe. These experimental results, when compared to the theoretical minimum required amplitudes (as described in equation 6) suggest that the probe has suitable versatility to counteract any surface force interactions during probing.

5   Conclusions

In order to understand the interactions between a vibrating micro-CMM probe and a measurement surface, a model to describe a driven oscillator subject to a damping force was developed. The surface forces that constitute the damping force were investigated and modelled and, from this, an expression for the minimum required amplitude of vibration was developed. This theoretical minimum was compared to experimental data collected using a laser Doppler vibrometer, which confirmed that the vibration of the probe is suitable to counteract the surface interaction forces.

6 Future Work

Work will continue on the development of the micro-CMM probe by comparing the theoretical amplitude and frequency shifts, calculated from equation 1, to real data collected from interaction experiments of the vibrating micro-CMM probe with a measurement surface.

Acknowledgments

This work was funded by the UK National Measurement Office Engineering Measurement Programme 2008 to 2011.

References

1. Leach, R.K.: Fundamental principles of engineering nanometrology. Elsevier, Amsterdam (2009)
2. Lewis, A.J.: A fully traceable miniature CMM with sub-micrometre uncertainty. In: Proc. SPIE, vol. 5190, pp. 265–276 (2003)
3. Weckenmann, A., Estler, T., Peggs, G., McMurty, D.: Probing systems in dimensional metrology. Ann. CIRP 54, 657–684 (2004)
4. Stoyanov, S., Bailey, C., Leach, R.K., Hughes, B., Wilson, A., O'Neill, W., Dorey, R.A., Shaw, C., Underhill, D., Almond, H.J.: Modelling and prototyping the conceptual design of 3D CMM micro-probe. In: Proc. 2nd ESITC, Greenwich, September 1-4, pp. 193–198 (2008)
5. Seugling, R.M., Darnell, I.M., Florando, J.N., Woody, S.C., Bauza, M., Smith, S.T.: Investigating scaling limits of a fibre based resonant probe for metrology applications. In: Proc. 12th ICPE, Portland, USA, October 19-24 (2008)
6. Lambert, P.: Capillary forces in microassembly. Springer, New York (2007)
7. Van Brussel, H., Peirs, J., Reynaerts, D., Delchambre, A., Reinhart, G., Roth, N., Weck, M., Zussman, E.: Assembly of microsystems. Ann. CIRP 49(2), 451–472 (2000)
8. Sitti, M., Hashimoto, H.: Force controlled pushing of nanoparticles: modelling and experiments. In: Proc. IEEE/ASME International Conference on Advanced Intelligent Mechatronics, Atlanta, September 19-23, pp. 13–20 (1999)
9. Andersson, K.M.: Aqueous processing of WC-Co powders. Doctoral Thesis, Royal Institute of Technology, Stockholm (2004)
10. Arai, F., Andou, D., Fukuda, T.: Adhesion forces reduction for micro manipulation based on microphysics. In: Proc. IEEE 9th Annual International Workshop on MEMS, San Diego, Febryary 11-15, pp. 354–359 (1996)
11. Lamoreaux, S.K.: Demonstration of the Casimir force in the 0.6 to 6 μm range. Phys. Rev. Lett. 78(1), 5–8 (1997)
12. Sattler, R., Wachutka, G.: Compact models for squeeze-film damping in the slip flow regime. NSTI-Nanotech. 2, 243–246 (2004)
13. Kaye, G.W., Laby, T.H.: Tables of physical and chemical constants, 16th edn. Longman, London (1995), http://www.kayelaby.npl.co.uk/ (2008)

PART III

Gripping and Feeding Solutions for Micro-assembly

Chapter 6

High Precision Positioning and Alignment Techniques

Alignment Procedures for Micro-optics

Matthias Mohaupt, Erik Beckert, Ramona Eberhardt, and Andreas Tünnermann

Fraunhofer Institute for Applied Optics and Precision Engineering
Albert-Einstein-Strasse 7, D-07745 Jena, Germany
Tel.: +49 3641 807 342
Fax: +49 3641 807 604
matthias.mohaupt@iof.fraunhofer.de

Abstract. The alignment procedure is an important step in the process chain of the assembly of micro-optical components and has a direct impact on the system performance of the micro-optical system, the necessary assembly time and the manufacturing costs. For these reasons, only alignment procedures that are adapted to the special requirements of each assembly task will be able to save costs and attain the best optical system performance. The paper describes the alignment methods and illustrates the assembly procedures based on selected micro-optical examples.

Keywords: Alignment, assembly, micro-optics, adhesive bonding, laser beam soldering.

1 Alignment Requirements of Micro-optics Assembly

The requirements of micro-optical alignment tasks are defined by the optical system design. The calculated tolerance budget describes the number of Degrees of Freedom (DOF) and the alignment accuracy. When a lens is placed into a mount, the rotational DOFs around the Rx- and Ry- Axis are defined by the mechanical stops of the mount, the z-DOF (along the optical axis) is also defined by the mount. The lens has to be centred into the mount with an accuracy of 10 µm to 50 µm to allow for manufacturing tolerances and the necessity of fixation with minimal mechanical stress applied to the optical parts, or to create an athermal mounting.

In comparison with a lens fixation, the DOF for the alignment of micro lens arrays often require a 6 axis alignment as a result of of the mounting accuracy of the CCD chip. Using the CCD socket as the fixation base, the CCD chip can be misaligned with respect to the socket by up to 100 µm in x- and y-direction and rotated in Rx and Ry direction by up to 200 mrad. Depending on the pixel size and the design of the micro lens array, the assembly requirements in x - and y-directions (overlay accuracy) are in a range of 0.1 µm to 1 µm. The micro lens array has to be rotated (Rz) against the CCD pixel lines with an accuracy of 5 µrad to 50 µrad. The alignment accuracy of the fibre to fibre coupling depends on the fibre core diameter and is in the range of 1 µm to 5 µm using multimode fibres and 0.05 to 0.1 µm using single mode fibres. If the fibres are also used for polarized application, the rotation around the optical axis also has to be aligned.

S. Ratchev (Ed.): IPAS 2010, IFIP AICT 315, pp. 143–150, 2010.

The alignment of the collimating lens is the most important assembly task during the manufacturing of laser diode modules. Most of the applications require a 6 DOF alignment of the lens to reach optimal focusing of the emitted beam profile with an alignment accuracy of 1 µm to 5 µm. Using moulded aspherical lenses, the alignment accuracy requirements can relaxed up to 50 µm.

The special charactertistics of using micro-optical components often require adapted handling and alignment devices due to the small part geometries, the low mass and the brittle material properties.

2 Alignment Methods

Two basic principles for the alignment of optical- and micro-optical systems can be used, active and passive alignment. Active alignment means that the optical system properties in function are used as a feedback for closed loop control.

During the alignment of a focusing lens with respect to a laser diode module, the beam profile in the far field and the detected intensity of the system are the criteria for the alignment status. During fibre coupling, the factor of optical damping is used for alignment. In special cases, wave front properties are measured by wave front sensors and used as alignment criteria. The detected intensity on the pixels of a CCD sensor is used for the alignment of micro lens arrays. By a calculation of the row and line intensity, the alignment can be precisely controlled. If every pixel of the CCD sensor detects the same intensity, the alignment can be finalized.

In contrast to to active alignment, a passive alignment procedure is based on the detection of geometrical properties of the components or alignment structures without any closed loop control by the optical system output. The passive alignment is done by using mechanical stops of the mount to orientate the optical components. Furthermore, the detection of outside diameter or edges of prisms is used for the alignment of micro optical parts with respect to each other. This kind of alignment is often supported by image processing for detection and calculation of the distances of the parts to each other and to the optical system. The technique of mark detection is also a passive alignment procedure. By using lithographic structured alignment marks on MEMS or MOEMS, accuracies of less than 1 µm can be attained.

3 Alignment Examples

3.1 Assembly of a Blue Ray DVD Pickup System

The hybrid optical system (fig. 1) works at a wavelength of 407 nm. The laser source of the pickup is a blue laser diode, which is soldered on a base made of thermal conductive material for reasons of thermal management. The generated laser beam is coupled into a prism assembly, reflected by a 90° prism to an objective lens and focused into the optical disc. A half wave plate (HWP) with a structured aperture for stray light reduction on it rotates the polarisation at 90°. The reflected light is focused into an arrangement of two detectors using a polarising beam splitter (PBS) layer which is deposited onto one prism. The polarisation vector of the laser beam is rotated

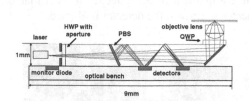

Fig. 1. Optical design of the DVD pickup   **Fig. 2.** HWP/QWP assembly device

in a range of 45° two times passing the quarter wave plate (QWP). A segmented pho-
todiode detects the high frequency data signal and can also be used for controlling the
track and focus determination.

At first, the laser diode is soldered to the substrate, whereby the direction of the laser beam is given by the orientation of the gap of the laser diode. After that, the
prisms are fixed by adhesive bonding at the prisms' surface on which a polarising
beam splitter layer is deposited. This alignment task is done using a precise assembly
device. The prisms are fixed by the assembly device; the orientation of the prisms to
each other is given by mechanical stops. The prisms are fixed to each other by adhesive bonding.

An aperture is structured onto the surface of the half wave plate (HWP), which is
assembled to the surface of the first prism, to define a special diameter of the optical
beam. The aperture has to be aligned with an accuracy of ± 25 μm perpendicular to
the optical axis (x-and y-direction). The assembly step is done using a device that has
mechanical stops for alignment of the HWP to the prism assembly (fig. 2). Before
curing the adhesive, the alignment status is checked by image processing which calculates the distance of the aperture to the centre of the prism assembly.

The prism assembly is aligned and fixed by adhesive bonding to the optical bench
in the next assembly step. The alignment of the prism assembly takes place with respect to the optical axis (beam) given by the assembled laser diode. The prism assembly is gripped by a vacuum gripper and placed onto the optical bench. After placing
onto the optical bench, the distance of the laser diode to the edge of the first prism is
measured using image processing. If the calculated distance is in a range of 20 μm
and the orientation of the prism assembly to the optical axis is in a range of 1°, the
prism assembly can be fixed by adhesive bonding.

The next alignment step is the mounting of the lens on the prism assembly. The lens is
fixed in a precisely manufactured mount prior to that and requires DOF for the x and y
directions (fig. 3), using a wave front measurement as the alignment criteria. The lens is
moved in x and y directions on the prism surface with a step size of 0.1 μm. After scanning, the lens is fixed to the prism assembly in the position of the lowest amount of wave
front error. By aligning the objective lens to the laser beam, the correction of further
assembly steps is possible. Finally, the assembly task for the alignment of the detector for
focusing control and data analysis is very important and sophisticated as a result of the
small detector geometry and the necessary contacting of the detector pads for signal
analysis. The DVD pickup system is switched on during the detector alignment; the focus

of the system is reflected by a moveable disc (fig. 4). Because of sinusoidal moving of the focus in the optical axis (z-direction) with an elongation of around 50 µm, the detected focus signal can be used for the active alignment of the detector. If the variable signals are symmetrical to the outer two detector segments, the detector is aligned.

Fig. 3. Lens alignment   **Fig. 4.** Detector alignment

3.2 Assembly of a Micro Lens Array on a CCD Sensor

The assembly and alignment of micro lens arrays (MLAs) is a necessary process step to increase the fill factor of special CCD sensors, e.g. in astronomical applications. To increase the intensity per pixel, the light of neighboured pixels is collected by a micro lens and focused onto one pixel. Linear arrangements of cylindrical lenses often collect the light of 3 or 4 pixel lines into one pixel line. Depending on the pixel size, the required alignment accuracy is less than 1 µm after alignment and fixation. To reach the best performance of CCD sensors and to eliminate the position tolerances of the CCD chip mounted into a ceramic socket in this application a 6 DOF alignment is necessary to align the spots of the micro lenses to the CCD sensor pixels.

The assembly concept requires a mounted MLA that is aligned to a frame. The frame is a structured holder made of ceramic or steel with a CTE that is adapted to the CCD socket material or to the silicon substrate of the CCD. The MLA is fixed to the frame by adhesive bonding using a two component epoxy adhesive. The mechanical stops integrated into the frame guarantee a pre pre-alignment of the MLA to the frame.

The first assembly step is gripping the MLA at the frame and positioning it with respect to the CCD. By detection of the pixel intensity, the alignment in z-direction (optical axis) is the first alignment step. After that, the rotation around the x- and y-axes is aligned by detecting the homogeneity of the intensity over the whole CCD pixel array.

The alignment of the rotation of the z-axis is the next important assembly step. The summarized lines and rows intensities are calculated and used for alignment of rotation around the z-axis for the MLA against the CCD sensor. The intensities of all pixels should be in the same value along every pixel line. The final assembly step is the alignment in x- and y-direction, meaning the alignment of the focus spots into the centre of the CCD sensor pixels. By comparison of neighbouring lines, the alignment information of the overlap accuracy of focusing the micro lenses onto the sensor pixels can be derived. Alignment steps of 0.5 µm can be detected by the camera control software. The positioning device used for alignment is a 6 DOF hexapod. The gripper holding the MLA frame is mounted onto the moveable platform of the positioning device. The user interface allows for defining a pivot point in the centre of the MLA, so the alignment can easily be controlled.

If the alignment is done, the gap between the CCD socket and the MLA holder is filled with a two component epoxy adhesive and the MLA is fixed to the CCD socket. By using an adhesive curing for 24 h at room temperature, minimal misalignments between the MLA and the CCD frame of less than 1 µm was measured.

Fig. 5. Assembly of an MLA on a CCD sensor

3.3 Alignment of a Fast Axis Collimator Lens to a Laser Diode

A further example for the active alignment can be given by cylindrical aspheric Fast Axis Collimator (FAC) lenses that need to be assembled in front of diode laser bars or single laser emitters. The usage of Bottom Tabs enables for all necessary six degrees of freedom for alignment while simultaneously creating virtually zero joining gaps between FAC and Bottom tab as well as Bottom Tab and heat sink of the diode or bar, which is the assembly reference in this case. Alignment and joining of this assembly

Fig. 6. Laser diode module with fixes FAC lens

is a complex process in which four solder bumps are applied overall, inter alia at the outer geometry of the FAC lens (fig. 6). It is evident that dealignment not only occurs due to application of the solder, but also due to overall handling, alignment and joining and the mechanical and thermomechanical stress introduced in the process. After basic optimisation of the process, the results indicate that it is possible to reach a dealignment of the FAC lens in the range of ±0,5 µm.

4 Fixation of the Alignment Status

After the alignment process, the fixation of the aligned micro-optical components has to secure the alignment status over the lifetime of the optical system. The fixation can not be done by mechanical clamping, known from macro-optics, for reasons of the small geometry. The mechanical fixation elements are one or more dimensions larger than the micro-optical system. Consequently, the fixation of micro optics is done by adhesive bonding or laser soldering.

4.1 Adhesive Bonding

Joining by adhesive bonding is the most common technology for the fixation of micro-optical components. The most important properties of optical adhesives are a high transmission in the spectral region of interest and the possibility of refractive index matching to the interfacing optical materials. To reach a minimal misalignment during curing, adhesives with minimal shrinkage (about 1-2 %) and well defined drop volumes as well as a symmetric placement of the drops with respect to the optics are necessary. Adhesives based on curing by UV-radiation mechanism were used for alignment tasks which require lengthy alignment procedures and also allow for very short process times during an automated assembly. Two component epoxy adhesives curing several hours at room temperature were used for joining components with very low mechanical and thermal stress generation during curing. These adhesives reach tensile strength up to 35 MPa and were successfully used in space applications.

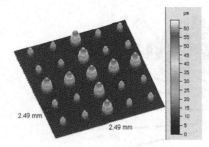

Volumes: 6pl and 120 pl

Fig. 7. Array of adhesive drops

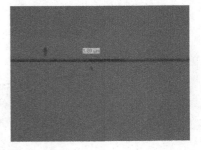

Fig. 8. Adhesive gap of < 3 µm

Due to an increasing miniaturisation of optical systems, the application of very small drops of adhesive, the realisation of well defined gaps filled with adhesive, and uniform adhesive layers between the optical elements are becoming important. The application of adhesive volumes down to the picolitre range can be realised using a micro-drop dosage system (figure 7).

The application of spherical spacer elements to the adhesive layer (micro pearls) can be used to obtain well defined gap geometries in the µm-range (figure 8). The well-known limitations of adhesives are operating temperatures lower than 120°C, long term stability, yellowing and poor environmental humidity resistance.

4.2 Laser Beam Soldering

To reach higher thermal and mechanical stabilities and to eliminate aging effects of the polymer adhesives, a soldering technology alternative has been adapted to micro-optical assembly. To use the soldering technology in optics, a metallisation of the optical components is necessary. A coating technology for the metallisation of optics and for depositing AuSn solder layers directly onto the optical components was developed and investigated /2/.

Fig. 9. Lens soldered by solder ball bumping (front and top view)

The solder ball bumping technology known from micro-electronics was successfully integrated into the micro-optical assembly process chain.

A misalignment during and after fixation by solder ball bumping in a range of less than 0.2 µm was attained /3, 4/.

5 Conclusion

The aspects of alignment and assembly of micro-optics were illustrated by examples using the active and passive alignment procedures. Using different fixation technologies, the alignment status can be secured with an accuracy of less than 1 µm. The handling devices and the environment condition have to be adapted to the special requirements of micro-optic alignment.

References

1. Mohaupt, M., et al.: Joining Procedures for a high precision assembly of micro-optical systems. In: 7th international conference and 9th annual general meeting of the European society for precision engineering and nanotechnology, Bremen: Conference Proceedings, vol. II, pp. 304–307 (2007) ISBN 10: 0-9553082-2-4
2. German patent application DE 10, 002 436.5 Verfahren zum Fügen justierter diskreter optischer Elemente (2007)
3. Beckert, E., et al.: Solder bumping - A flexible joining approach for the precision assembly of optoelectronical systems. In: Conference Information: 4th International Precision Assembly Seminar, Chamonix, France, February 10-13. Micro-Assembly Technologies and Applications, vol. 260, pp. 139–147 (2008)
4. Burkhardt, T., et al.: Parametric investigation of solder bumping for assembly of optical components. In: Laser-based Micro- and Nanopackaging and Assembly III. Proceedings of SPIE, vol. 7202, p. 720203. SPIE, Bellingham (2009)

Pneumatic Driven Positioning and Alignment System for the Assembly of Hybrid Microsystems

Christian Brecher, Martin Freundt, and Christian Wenzel

Fraunhofer Institute for Production Technology, Steinbachstrasse 17,
52074 Aachen, Germany, Tel.: +49(0)241-8904-253, Fax: +49(0)241-8904-6253
martin.freundt@ipt.fraunhofer.de

Abstract. Micro assembly is typically characterized by positioning tolerances below a few micrometers. In the case of the assembly of hybrid micro systems, such as optical systems, micro ball lenses or micro probes for measurement tasks, even positioning accuracies in the sub-micrometer range have to be achieved. The efficiency of the use of automated handling devices is strongly influenced by the flexibility of the equipment and the required application specific customizations. In this context a high precision assembly head is presented. It upgrades conventional robots with the ability to do fine alignment steps in sub-micrometer resolution and 6 DOF. Therefore it is equipped with a universal endeffector structure.

Keywords: Optical assembly, Alignment, Micro Assembly, Micro Gripper, Assembly Head, Positioning, Air Bearing, Pneumatic Actuator, FlexibilityKeywords.

1 Introduction

Micro assembly is typically characterized by positioning tolerances below a few micrometers [1]. In the case of the assembly of hybrid micro systems, such as optical systems, micro ball lenses or micro probes for measurement tasks, even positioning accuracies in the sub-micrometer range have to be achieved [2].

The assembly is one of the most value creating process steps, as it allows to combine two materials or production processes in one product. Especially for hybrid micro systems, the assembly process step creates product functionality respectively a product value, larger than the total of the two assembled components had before. This merging of technological properties takes place within electro optical systems and other advanced technology focused applications.

Due to the need of highly accurate assembly systems and extensive alignment procedures, the assembly of high precision optics and micro systems is still characterized by customized solutions [3, 4]. On behalf of this background, the Fraunhofer IPT has developed a new approach how to realize a highly flexible, fast and cost-efficient hybrid micro assembly processes. Main element of this approach is a robot guided assembly head, capable of adding high precision manipulation capabilities on conventional handling devices.

Within a handling process of the hybrid system, the robot guided active assembly head will be pre-positioned by a conventional handling device like an industrial robot.

S. Ratchev (Ed.): IPAS 2010, IFIP AICT 315, pp. 151–158, 2010.
© IFIP International Federation for Information Processing 2010

Thereby the large working area and high dynamics of the conventional system can be combined with the features of the assembly head, designed to compensate position and alignment errors with sub-micrometer accuracy. The manipulation precision required for the micro-assembly is therefore subsequently realized directly at the tip of the assembly head.

1.1 Process Scenario and Requirements

Regarding universal usability, which has to be a goal, designing an assembly head which is supposed to be used as a all purpose tool, adding precision alignment functionalities to a conventional robot system, following challenges have to be considered:

1. **Universal endeffector:** Design of an endeffector or endeffector system capable to be used for various part materials and shapes. Therefore it should be able to compensate for process or part tolerances.
2. **6 DOF fine alignment:** A fine alignment stage capable of compensating the position deviation of a conventional robot in all, 6 degrees of freedom. It also has to be capable to allow for a certain additional high precision manipulation movement after the deviation compensation process.
3. **Referencing Sensor system:** A solution allowing the assembly head to be used not only within active guided assembly processes but be also an adequate tool for open loop positioning processes.
4. **Ease of integration:** The assembly head, containing and providing the precision specific functionalities to the assembly process has to be designed in a way that allows for standard integration procedures. Therefore the assembly head has to come with its own control so that it can be used as a plug and play device, only requiring power, digital control signals a compressed air supply.

These boundary conditions result in a process chain, described in figure 1. Within this process the conventional robot is used for prepositioning with its specific dynamic and precision. Once the robot has reached its designated position, the assembly head starts with the high precision manipulation movement. In case of an open loop positioning process, the assembly head therefore acquires its position relative to a reference mark localized next to the working place with its integrated sensor system. In case of an active guided assembly and alignment process the assembly head can be utilized to directly adjust the part. For the handling and alignment of optics, especially in the field of the diode laser fabrication, the parts have to be positioned within increments of about 100 µm respect angle manipulation increments in the range of

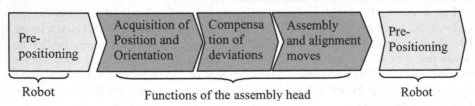

Robot   Functions of the assembly head   Robot

Fig. 1. Component handling process of a hybrid positioning device consisting of conventional robot and active assembly head

Table 1. Essential requirements for a fine alignment system integrated into an assembly head for the realization of high precision manipulation

Function	Requirements
Alignment	6 DOF, 0.1 µm / 0.001° increments, 1 mm / 2° travel range
Integration issues	Max. workspace about 150x150x150 mm³, cone shaped outline for a good accessibility to the part, small mass, robust about acceleration, miniaturized
Process requirements	Flexible usable gripper tool, usable for various part shapes, does not require endeffector exchange

1/1000°. The typically actively controlled adjustment process requires alignment in up to 6 DOF in order to allow for the product quality optimisation.

Concerning the combination of a conventional robot and an assembly head, as well requirements of the conventional robot system have to be considered. As conventional robots are not capable to reach the required precision demands, the drives of the conventional robot have to be deactivated in order to suppress control circuit vibrations. A survey of the main requirements is listed in Table 1.

In this context, an assembly head was developed which includes a 6 DOF fine alignment system, a universal endeffector and a referencing sensor system all controlled by an integrated control circuit allowing the use of the system as exchangeable gripping device, empowering conventional positioning devices to be applied for high precision assembly processes.

2 Design of the Universal Usable Endeffector with Alignment Capabilities

The design of the manipulation and alignment system has to meet the requirements of the endeffector working principle, as it is the system component carried by the rest of the system. Exchange and load scenarios, stiffness and precision requirements characterize the resulting requirements for the alignment system. The design challenge is to combine both functionalities within one device. Therefore, a concept of a hybrid endeffector and alignment structure was designed, allowing gripping and manipulating various materials and part shapes. The structure bases on three identical 3 DOF devices, each positioning one endeffector in X, Y and Z direction, as shown in figure 2. As the distance between the three endeffectors towards each other has a significant influence on the manipulation range and resolution, the orientation of the X-axis of all stages is radial to the centre of the structure as shown in figure 4. As the main axis, it is driven by a combination of sequential controlled elliptec piezomotors, is capable to travel a significant distance of about 12 mm. The second axis is orientated within tangential orientation, thereby. The third axes, in Z-direction allows for lifting and lowering the component as well as tilting the component around X and Y-axes. Both, the second axes and the third are capable of travelling 1-2 mm depending selected resolution.

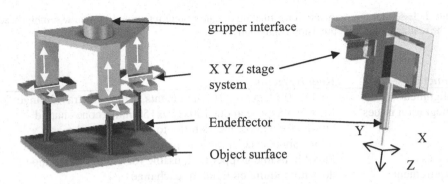

Fig. 2. Illustration of the function of a highly flexible and precise fine alignment device, system structure (left), function principle of the three axis stage (right)

Thereby following challenges within the design of robot guided, flexible and high precision assembly heads are addressed:

Flexible endeffector function. Gripping the component in three contact points creates a shape independent endeffector function. The three end effectors have to be designed in a way that they cannot apply any serious torque to the component. Thereby they define the position and orientation of the component and ensure a distinct handling process.

6D alignment functionality. Each of those end effectors, defining the position and orientation of the gripped component can be positioned within three translational degrees of freedom. With the adequate control system, any part shape can be manipulated in all degrees of freedom. The concept of three separate end effectors, allows to reduce the technical complexity. The design challenge of the fine alignment system can be reduced from a 6 DOF system towards a comparable simple 3 DOF system. This design concept is the key to be able to design a high precise and same time miniaturized manipulation device of six degrees of freedom.

Integration into an automation environment. The assembly head includes all required actuators, sensors and control devices, allowing the integration into existing robot systems without extensive customizing of the automation setup. It will have to be supplied with compressed air, power and the digital data for the execution of gripping and alignment tasks via a SPI-bus-interface. This ensures the usability of the system and allows a tool like use for various precision handling and alignment tasks.

3 Design of an Endeffektor Manipulation Stage

Due to the endeffector concept, the complexity of the design challenge is reduced to a 3 DOF positioning system. The requirements onto the device concerning stiffness and load capacity are also minimized, as only three axes have to be realized. Further, the components, going to be manipulated, are lightweight compared to required stage structures. An additional load introduced by a complex endeffector can also be spared.

Concerning high precision requirements, all guidings are based on air bearings. Those air bearings are integrated within the structure of the slides due to construction space and lever minimisation. The three required 3 DOF positioning devices are equipped with miniaturized air bearings to ensure absolute precision capabilities. Its main slide is driven by a pattern of six elliptic piezo motors allowing to archive long travel ranges of more than 10 mm. For high precision movements, an optical sensor system is integrated within the main stage, allowing to measure the primary slide position. Slide two and three are integrated within the structure to optimise the load profile on the air bearings and reduce the size of the stage system. Figure 3 illustrates the integral design. In order to avoid disturbing forces caused by tubings for the supply of the air bearings and wiring for actuators, a system of contact free transmissions for compressed air is integrated within the air bearing gabs. (patent application DE 10 2007 023 516 A1).

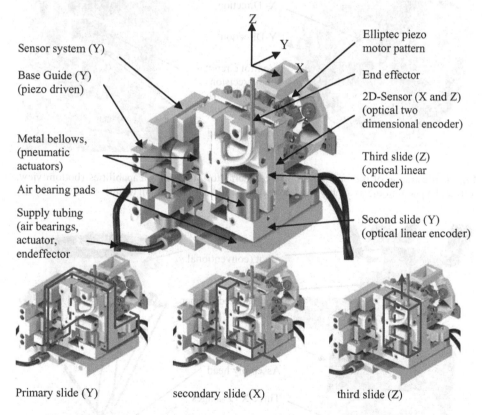

Fig. 3. CAD-Model of the three axis, air bearing stage with internal transmission of compressed air and pneumatic driven second and third axis

They allow for supplying of the stages and for providing the compressed air signal to the metal bellows actuators, driving the secondary and third stage. A 2D planar encoder device integrated on the first slide detects the position of the second and third slide. It allows for the accurate measurement and control of the second and third axis, which are driven by the comparable weak but extremely compact metal bellow actuators. Due to the complex air bearing structure and the tight tolerances within air gab, all air bearings are designed preloaded by adjustable magnets. This adjustment opportunity of the preload enables to put the complex air bearing design into operation. Due to the required design with several air bearing pads, integrated within the slides all slides are fabricated using laser sintering technology allowing to route free form boreholes and cavities through the slides and thereby enable to supply each single air bearing pad, made out of porous graphite. All air bearing surfaces are ultra precision diamond milled allowing for small bearing gabs and a precise alignment of the slides towards each other.

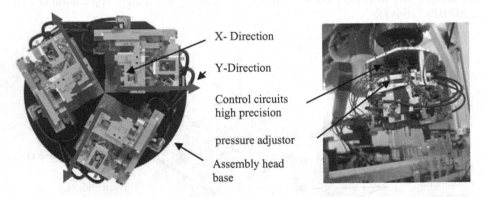

X- Direction

Y-Direction

Control circuits
high precision

pressure adjustor

Assembly head
base

Fig. 4. Illustration of the orientation of the endeffector positioning capabilities (bottom view, left) and a photo of the assembly head prototype within a robot system (right)

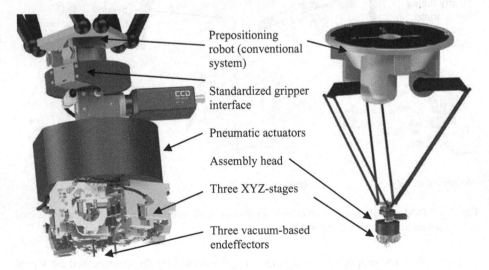

Prepositioning
robot (conventional
system)

Standardized gripper
interface

Pneumatic actuators

Assembly head

Three XYZ-stages

Three vacuum-based
endeffectors

Fig. 5. CAD-Model of the assembly head based on three identical XYZ-stages each positioning a vacuum based, leakage afflicted endeffector

The three stages are oriented under 120 degree. The main slide thereby is orientated in radial direction allowing an adaption of the end effector configuration. In order to adapt to component shape and size, the distance of the end effectors can be modified from 0.2 mm to about 12 mm. The assembly head is equipped with an on board electronic including sensor evaluation circuits, microprocessor based control unit and driver circuits for the elliptic piezomotors. The whole assembly head thereby requires compressed air to run the air bearing and actuators, 24 V power. The communication with a superior process control is avaible via RS232 or SPI. This ensures that the assembly head can be used as an all-purpose tool within conventional robots. Figure 5 illustrates the integration within a robot system at the Fraunhofer IPT. With this assembly head design conventional robots will be usable for precision assembly processes. In case of objects which can not actively aligned in within a process, a referencing sensor device is required. In order to avoid part specific adaption work within process implementation, it is required to apply a universal sensor system. The sensor system has to be able to be integrated within the assembly head structure allowing using it in multiple positions within the assembly area.

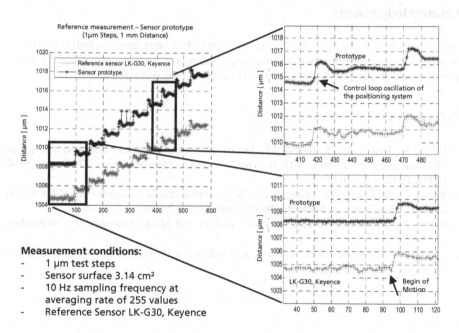

Fig. 6. Preliminary test results of a capacitive sensor element with sensor head integrated digital readout, designed for the setup of a 6 DOF robot guided referencing sensor system

Commercially one-dimensional sensor systems with adequate resolution in the sub micrometer range and even measurement range of more than 1 mm are avaible. In order to combine six sensors, allowing to measure the position and orientation deviation of the robot system, the signal analysis circuits have to be integrated within the assembly head. Due to the boundary conditions, especially the application within conventional robots, only digital signals can be failure-free transferred through a gripper interface and robot wiring. In behalf of this challenge a prototype of a compact capacitive sensor element with onboard signal digitalization was designed.

Preliminary testing of noise and resolution allows to expect detection resolution of about 100 nm. Figure 6 presents preliminary test results of a prototype sensor element which has a build in digital readout using SPI communication.

4 Outlook

The technology of gripping and manipulation components based on three vacuum based endeffectors presented here was implemented within a prototype assembly head. It is going to be put into operation and will be tested within its handling precision and usability. Additional research will be conducted within the field of sensor integration in order to be able to archive adaptivity and allow for the design of semi automated process programming.

Acknowledgements

The authors would like to thank the »Deutschen Forschungsgemeinschaft (DFG)« for supporting the »Greifer und Montagemaschinen« research project as part of the special research topic SFB 440 »Assembly of hybrid micro systems«.

References

1. Heuer, K., Hesselbach, J., Berndt, M., Tutsch, R.: Sensorgeführtes Montagesystem für die Mikromontage. In: Robotik VDI-Berichte Nr. 1841, München, June 17-18, pp. 39–46 (2004) ISBN 3-18-091841-1
2. Brand, U., Wilkening, G.: Dimensionelle Messtechnik und Normung für die Mikrosystemtechnik – Eine Frage des Maßstabs. In Mikroproduktion 3/2006
3. Raatz, A., Hesselbach, J.: High-Precision Robots and Micro Assembly. In: Proceedings CoMa (2007)
4. Peschke, C.: Mehr-Achs-Mikrogreifer zur Handhabung biegeschlaffer Mikrobauteile, Dissertation, RWTH Aachen (2007)

Flexure-Based 6-Axis Alignment Module for Automated Laser Assembly

Christian Brecher, Nicolas Pyschny, and Jan Behrens

Fraunhofer Institute for Production Technology IPT,
Department for Production Machines,
Steinbachstrasse 17, 52074 Aachen, Germany
nicolas.pyschny@ipt.fraunhofer.de

Abstract. Fully automating the assembly of laser systems puts high demands on the accuracy, but alike on the flexibility and adaptivity of the assembly system. In this paper a concept for a flexible robot-based precision assembly is introduced. This concept is based on a modular 6-axis alignment tool which has been designed as a hybrid serial-parallel manipulator with flexures for all revolute and spherical joints. Technical insights on the design and dimensioning based on analytical calculations will be presented as well as first results from the characterization of a prototypal alignment module.

Keywords: Precision assembly, parallel manipulator, compliant mechanism, flexures.

1 Introduction

Laser applications, especially laser marking and engraving, are about to reach the long desired status of a commodity. Thus, the technological developments of opto-mechanical products are driven by dynamically changing customer demands and miniaturization, posing a major challenge for high precision manufacturing and assembly [1].

Mainly, the assembly of optical systems is a very challenging task due to highest requirements on alignment accuracies and significant influences of component tolerances. Hence, nowadays the assembly is dominated by manual operations, involving elaborate alignment by means of adjustable mountings and the application of multiple sensors. Economically, further automated assembly of high-quality optical systems, such as laser units, could be the way to produce more reliable units at lower cost. Nevertheless, the multi-functional and hybrid character of lasers puts a high demand on assembly systems, which must be able to handle precision assembly and simultaneously cope with an increasing number of product variants.

Presently, a lack of suitable assembly systems with the required precision and flexibility prevents a time and cost efficient automated assembly. The state of the art is characterized by manual and semi-automated solutions with specialized joining technologies for different parts and components [2]. Existing automation solutions for precision assembly of hybrid products are rigid and inflexible and therefore highly

S. Ratchev (Ed.): IPAS 2010, IFIP AICT 315, pp. 159–166, 2010.
© IFIP International Federation for Information Processing 2010

restricted in their adoption for different assembly tasks. There is a lack of adequate equipment for flexible, automated and high precision gripping, manipulating, positioning, alignment and joining of optical components and systems [3].

2 Flexibly Automated Assembly of a Miniaturized Solid State Laser

New solutions for a flexible and automated assembly of hybrid high-precision products – focused on the assembly of a miniaturized diode-pumped solid-state (DPSS) laser for labeling and marking applications – are under development within the major research initiative "Aachen House of Integrative Production" [4]. The miniaturized laser system, designed and developed at the Fraunhofer ILT, is a DPSS laser in planar configuration enabling automated assembly. All optical components can be positioned and assembled from above and are soldered on a ceramic base plate (Figure 1). The approach to automated assembly is based on a thoroughly modular assembly cell, involving three industrial robots, equipped with configurable tools and sensors for handling and joining, machine vision and illumination. The entire system is continuously updated with information about the assembly process, provided from multiple sensors: not only the robot-based camera, but also a camera-based laser beam analysis system and additional sensors for controlling the laser itself.

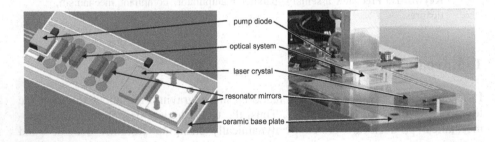

Fig. 1. Miniaturized solid state laser for automated assembly *MicroSlab[2]*: left shows the model of the optical system (*size of the base plate around 40mm x 100 mm*), right shows a scene from an assembly sequence

For the alignment of components in the miniaturized solid state laser system a flexible and fast handling system has been developed, consisting of a conventional industrial robot and a high-precision alignment module. The module is mounted to the imprecise but dynamic robot serving as a mobile six-axis alignment tool. It is pre-positioned within the large working area of the robot and compensates for position and alignment errors with sub-micrometer accuracy. For the assembly of the optical system and the resonator mirrors an active alignment based on the evaluation of the components' beam transfer functions will be applied (Figure 2).

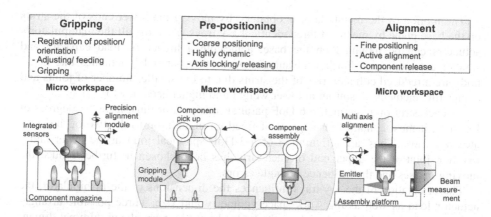

Fig. 2. Alignment concept with modular 6-axis fine-positioning module

A second robot carries modular sensor tools for the evaluation of alignment results, and an innovative soldering technique is being developed and implemented in another robot-based tool to allow for a fully-automated assembly.

3 Kinematical Analysis and Optimization of Alignment Module

For the precision manipulation of optical components a miniaturized alignment module with six degrees of freedom (DoF) was to be developed that provides high resolution and high stiffness for robot-based alignment operations. Technical requirements for the laser assembly were an overall weight of less than 1kg with outer dimension of about 100 mm x 100 mm x 100 mm, a payload of 200 grams and sub-micrometer resolution in a work-space of at least 1 mm x 1 mm x 1 mm. Required tip and tilt angles for the alignment of lenses and mirrors had been specified as ±0,3°.

For the kinematics of the six-axis manipulator a symmetric, hybrid serial-parallel structure has been chosen that consist of three inextensible struts which connect three non-collinear points of its platform to its base (Figure 3).

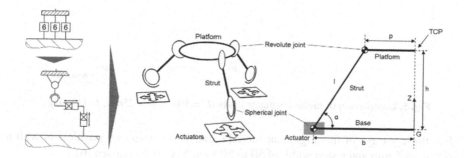

Fig. 3. Kinematical structure of the alignment module: three parallel kinematic chains with six DoF each *(left top)*, consisting of an actuated serial x-y-tray, a spherical joint and a revolute joint (left bottom); base radius b, platform radius p, length of the struts l, strut angle α, height h

The motion of the manipulator is obtained by moving the lower ends of the struts on the base plane by means of three identical x-y trays. Thereby, all the manipulator's actuators can be mounted on the base, achieving a higher payload capacity and smaller actuator sizes. Further advantages over other parallel kinematics like hexapods are a reduced collision risk of the struts due to the smaller number of joints and connecting elements, resulting in lower weight and higher accuracy [5].

As each strut of a symmetric 6-DoF parallel manipulator must have six degrees of freedom, it must contain four degrees of freedom in addition to the two of the actuated x-y trays. Thus, a configuration of a 3-DoF spherical joint at the lower and a revolute joint at the upper end of each strut has been chosen to further reduce the moving masses and the influence of angle errors.

The orientation of the x-y-trays influences the dimensions of the manipulator, the actuator load as well as the work space geometry and size. Three possible arrangements are depicted in figure 4 where the table shows the normalized minimal dimensions of the manipulator base. For type 1 all trays are aligned with the x- and y-axis of the base coordinate system. This leads to a minimal required space and a rectangular workspace in each horizontal section, while type 2 and 3 lead to workspaces with 120° symmetries, in line with the kinematical design.

		Type 1	Type 2	Type 3
Base diameter [%]		100	118	107
Outer diameter [%]		204	207	212

Fig. 4. Possible orientations of the x-y-trays

A vertical payload (z-axis) is most evenly distributed on the actuators for type 3 orientation, reducing the required actuating and blocking forces [Figure 5].

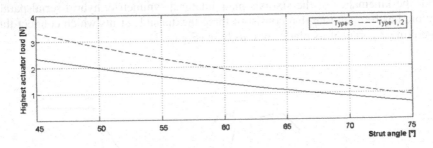

Fig. 5. Comparison of highest actuator loads (*b = 50mm, p = 25mm, load = 1kg*)

For the x-y-trays of the alignment module stacked piezo-based positioners with a travel of ±5.5 mm and a step width of 50 to 500 nm have been chosen [6].

These linear positioners (SLC-1720) exhibit a very good ratio between outer dimensions and travel range with actuating forces above 2 N, blocking forces above 3 N and allow for vertical loads of 40 N.

The resulting workspace of the kinematics are shown in figure 6 for the different types of actuator orientation. As mentioned above a symmetrical design of the manipulator leads to a workspace that is symmetrical to the y-z plane as shown.

For work space analysis the biggest possible cube inside the workspace is being identified, where the platform's zero position forms the center of the cube. One-dimensional comparisons are based on the edge length of this cube – called work space index. A calculation of work space indices for all actuator orientations shows an advantage for type 3 of 4,992 mm compared to 4,372 mm for type 1 and 2.

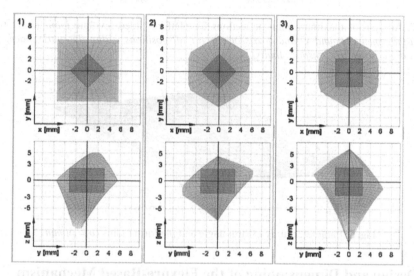

Fig. 6. Highest actuator loads *(b = 50mm, p = 25mm, α = 45°)*

For the optimization of the geometrical dimensions, an analysis of the central stiffness matrix allows the derivation of following design guideline: If the platform radius is half the size of the base radius, the stiffness matrix will be diagonalized, i.e. the deformations induced by applied forces and moments are decoupled. Applied forces only cause translational, moments only rotational deformations [5].

Other aspects that have been analyzed and taken into account are the size and shape of the resulting workspace, the overall size of the system as well as the maximum deflection of the spherical and revolute joints. The latter aspect is of special relevance for the design of flexure-based mechanisms as the required range of motion is the decisive parameter for the dimensioning of flexure joints. Figure 7 shows an exemplary analysis of maximum joint deflection for the platform-sided revolute joints. The underlying calculations are based on tip-tilt movements around the x- and y-axis of ±3° for all positions within the workspace.

Figure 8 depicts the effect of changes in platform radius and strut angle on the size of the work space, represented by the introduced work space index. With a defined base radius the influence of the strut angle proves to be greater, as the transmission ratio of vertical platform travel (z-axis) to actuator travel is primarily depending on this parameter, i.e. the smaller the strut angle, the bigger the transmission ratio and thereby the larger the vertical travel range of the platform.

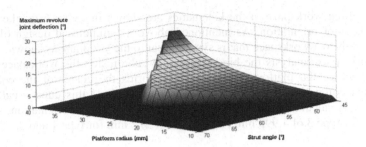

Fig. 7. Analysis of maximum joint deflections for platform sided revolute *joints (b = 50mm, tip-tilt-movements around x- and y-axis of ±3° for all positions within the work space; zero values are a result of workspace limits)*

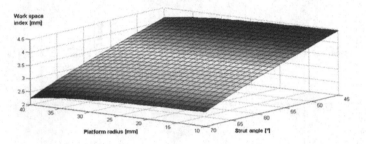

Fig. 8. Parameter study of strut angle and platform radius *(b = 50mm)*

4   Design and Dimensioning of the Flexure-Based Mechanism

To maximize the accuracy of the alignment module a continuous monolithic construction has been realized by designing all joints as flexures to eliminate the presence of friction, wear and clearances.

The developed analysis tools have been applied to calculate the required joint displacements as ± 8-10° for the tilt angles of the revolute and spherical joints and ± 12.5° for the torsion angle around the strut axis. For such large displacements a special type of flexure design has been chosen that differs from the wide-spread and well-known notch-type joints or leaf springs. These two groups of flexures have been researched for years and are widely used for small-displacement mechanisms with high requirements on precision, but suffer from poor off-axis stiffness and significant axis drift for larger deformations.

The applied concept for a compliant revolute joint has been presented in a paper from Kota and Moon [7]. The proposed design generates pure rotational motion with widely reduced axis drift as well as a superior off-axis stiffness compared to many other types of flexure designs. Joint deflections, i.e. rotational movements, are based on the torsion of a cross-shaped beam which results in off-axis stiffness that is 220 to 38.000 times higher than the stiffness around the axis of motion.

Based on this principle a miniaturized spherical joint has been designed for the manipulator. The three motion axes have been divided into two intersecting cross-shaped beams for the tip and tilt movements of the struts and a third flexure element for the torsional degree of freedom. The platform-sided revolute joints have been based on the same concept (figure 9).

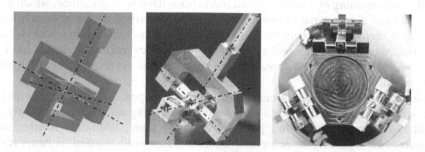

Fig. 9. Flexure design for spherical and revolute *joints (left: concept of spherical joint; middle: spherical joint designed for and manufactured by micro milling; right: top view of alignment module showing the platform-sided revolute joints)*

Based on the described design approach a first prototype of the manipulator has been set up. The calculated and simulated deformations of the flexures could be verified and measurements to characterize the accuracy and repeatability of the system have been conducted on a coordinate measurement machine (CMM).

The positioning accuracy of the manipulator could by means of kinematic calibration be improved from 50 micrometers down to 6 micrometers positioning accuracy within the biggest possible cube inside the workspace. An evaluation of the manipulator's repeatability showed that the measurements varied within the uncertainty of the CMM (around 0,75 micrometers). As the full potential of the design approach could not be verified by these measurements, first interferometer measurements have been conducted to characterize the motion resolution of the alignment module (figure 10).

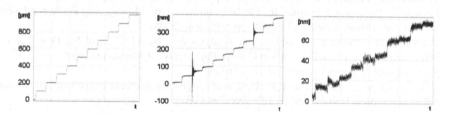

Fig. 10. Interferometer measurements to characterize the motion resolution (left: 100 μm steps; middle: 50nm steps; right: 10nm steps; closed-loop control of the actuators)

These measurements proved sub-micron motion resolution and promise more reliable results for a further characterization of accuracy and repeatability. It can be seen that the flexure-based mechanical design of the alignment module allows a widely unaffected transformation of the actuators' motion onto the platform

5 Conclusion and Outlook

Based on the requirements from producing miniaturized laser systems, a concept for a flexible robot-based precision assembly has been introduced that features a modular 6-axis alignment module. For a chosen hybrid kinematical structure different analyses for the dimensioning of a flexure-based mechanism have been presented, which led to the design of a prototypal alignment module. The described design approach makes use of an innovative concept for large-displacement compliant joints with superior qualities compared to other notch-type or leaf spring flexures. This concept has been adopted and extended to be applied for all spherical and revolute joints of the alignment module.

First measurement results from a CMM proved a repeatability of the manipulator below one micrometer, helped to improve the positioning accuracy significantly, but could not characterize the full potential of the system. Additional interferometer measurements regarding the achievable minimum step width showed far better results and will be continued for a more detailed characterisation of the alignment module.

Acknowledgments

The authors thank the »Deutsche Forschungsgemeinschaft (DFG)« for the support of the project within the Cluster of Excellence »Integrative Production Technologies for high-wage Countries« at RWTH Aachen.

References

1. Fatikow, S., Rembold, U.: Microsystem technology and microrobotics. Teubner, Stuttgart, Leipzig, p. 27 (2000)
2. Brenner, K.-H., Jahns, J.: Microoptics - From Technology to Applications, p. S. 34. Springer, New York (2004)
3. Tummala, R.: Fundamentals of microsystems packaging, New York, p. S. 18 (2001)
4. Loosen, P., Brecher, C., Schmitt, R.: Flexibel automatisierte Montage von Festkörperlasern, wt Werkstatttechnik online 11/12-2008, pp. 955–960 (2008)
5. Tahmasebi, F.: Kinematic Synthesis and Analysis of a Novel Class of Six-DOF Parallel Minimanipulators. Ph.D. Dissertation College Park: Univ. of Maryland (1993)
6. SmarAct GmbH, http://www.smaract.de
7. Kota, S., Moon, Y.-M., Trease, B.P.: Design of Large-Displacement Compliant Joints. Journal of Mechanical Deisgn, ASME 127, 788–798 (2005)

Approach for the 3D-Alignment in Micro- and Nano-scale Assembly Processes

Thomas Wich[1,2], Christian Stolle[1], Manuel Mikczinski[2], and Sergej Fatikow[1,2]

[1] University of Oldenburg, Division Microrobotics and Control Engineering,
D-26111 Oldenburg, Germany
[2] OFFIS Institute for Information Technology, Technology Cluster Automated
Nanohandling, Escherweg 2, D-26121 Oldenburg, Germany
{Thomas.Wich,Christian.Stolle,Manuel.Mikczinski,
Fatikow}@Uni-Oldenburg.de

Abstract. Most assembly processes on the nano-scale take place in a Scanning Electron Microscope (SEM) for the reason of high magnification range of the microscope itself. Like all microscopes, the SEM delivers visual data just in two dimensions. This is a bottleneck for all assembly processes which require of course information of the parts to join in a third dimension. This paper shows an approach with a dedicated sensor. As an example for an assembly process a carbon nano tube (CNT) is fixed on a sharp metal tip. The sensor used detects contact between these two parts by exciting a bimorph cantilever made from piezoelectric material. It is shown that with this approach the contact is reliably detected. Recent experiments on introducing a new excitation structure show the possibility to add more dimensional testing in the same way as the one dimensional type.

Keywords: Assembly process, carbon nano tube, TouchDown sensor, automation.

1 Introduction

The deterministic assembly of parts and components on the micro- and nano-scale is derived from assembly processes as known from the macro- and meso-scale. Although a lot of similarities exist for both regimes, micro- and nano-scale assembly faces some distinct differences, making it a challenging task both for manual and automated execution [8], [2].

Fig. 1 shows a typical flowchart for such an assembly process, which consist of the tasks *Transport*, *Adjustment*, *Joining*, *Separation* and *Inspection*. For the assembly of one product, these tasks have to be repeated in a sequence. In contrary to self- and parallel assembly, usually these tasks are not parallelised, but repeated n times for the production of n products.

The transport of the parts consumed during assembly usually involves several orders of magnitude in geometrical dimensions, e.g. actuator travel in the millimetre regime to a final position with nanometre accuracy is involved. This is mainly due to the fact, that micro- or nano-scale parts are often mounted on meso-scale parts. In automated assembly, these are fed to the assembly machine on adaptors [11], which

S. Ratchev (Ed.): IPAS 2010, IFIP AICT 315, pp. 167–173, 2010.
© IFIP International Federation for Information Processing 2010

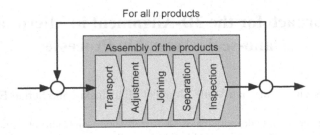

Fig. 1. Process layout for the automated serial assembly on the nano-scale

have dimensions controllable by human operators. In the adjustment task, two parts or even a part and the handling tool (e.g. a gripper) are adjusted relative to each other in position and orientation. This has to be done in six dimensions, however, depending on the part to be handled, some simplifications can be used (e.g. for spherical objects). However, the most important outcome of this task is the defined contact between the parts, which is a key prerequisite for the forthcoming joining. The joining task is the key task within the process chain, as two parts are united to one product here. With respect to micro- and nano-scale assembly, the number of available techniques is very limited, simply because not all macro- and meso-scale techniques can be down-scaled in a straight forward way. On the micro- and nano-scale, parts tend to stick to each other [5], to the underlying substrate and to the handling tools due to adhesive surface forces. To enable the automation of parts handling, they need to be stored on the adaptor in a well-defined state, i.e. with defined distances in between, separately and held in place. On the nano-scale, the latter one is e.g. achieved by growing parts on a substrate [7], resulting in a strong chemical or physical bond in between. Thus, for cutting the parts from the substrate or even separating them from the handling tool, a dedicated separation task is necessary. The inspection task serves as a measure of quality of the assembly process and usually involves testing of the bond between the parts.

Especially in the transportation and in the adjustment tasks, the 3D-alignment of parts or tool and part, respectively, is challenging for the above mentioned reasons. In section 2, these challenges will be described in more detail, resulting from the characteristics of the magnifying vision sensors. Based on typical sceneries in nano-scale assembly, the methods for in- and out of-plane coarse alignment will be discussed shortly. Additionally, the implemented approach for coarse alignment is discussed in more detail. Section 3 describes the TouchDown-sensor, a tool used for contact detection between tool and substrate or part, respectively. The results of the implemented alignment process will be described section 4 in more detail. Section 5 concludes the approach.

2   Challenges and Approaches in Aligning Tool and Object

Magnifying vision sensors are literally spoken the gate for accessing the micro- and nano-scale. Most commonly, optical light microscopes are used for the micro-scale; for accessing the nano-scale, scanning electron microscopes (SEM) and its derivates or - less frequently - scanning probe microscopes are used. Within this paper, the focus will be on SEM [1], [10]. Although they provide practical image acquisition speed, high and adjustable magnification, the high depth of focus and parasitic effects like image drift limit their applicability as a precise and unique global sensor [6].

The high depth of focus, usually considered as a benefit for analytical purposes, requires an experienced operator in manual assembly and dedicated sensor tools or methods in automated assembly, respectively, for out of plane alignment of part and tool. The parasitic effects, resulting from hysteresis effects on the scanning coils of the SEM and from charging effects in the scene, complicate the in-plane alignment. Therefore, the approach between tool and object is realised using *visual servoing* [9], [4], i.e. the distance between both objects is measured with object recognition methods in the recorded image. These relative measurements allow for precise alignment, however, the procedure requires continuous image acquisition, which limits the process speed compared to *open-loop*-alignment based on a world model.

The out-of-plane alignment is achieved by a combination of two methods. For coarse alignment, a series of images is taken at different focus distance [3]. This focus series is evaluated for sharpness in regions of interests (RoI), drawn around the tool and the part, resulting in two distinct curves. By comparing the sharpness-over-focus curves, the distance between tool and object is evaluated by the distance between the sharpness peaks for both RoIs. However, due to the high depth-of-focus of the SEM, the precision of this method is limited to several μm in the necessary magnification. Consequently, for final alignment, the tool and object are adjusted to each other until they are in mechanical contact. The contact is measured by the TouchDown-sensor. As an application example, we will follow the in- and out-of plane alignment sequence between a CNT and a fine etched metal tip further. Fig. 2 shows the starting scenery.

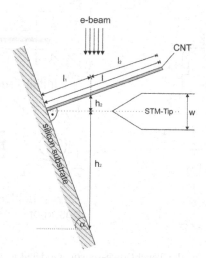

Fig. 2. Side view of the scene for the assembly of CNTs. The CNT is grown on a silicon substrate, which is mounted with a tilt angle in the SEM. The CNT is approached by a fine etched metal tip. A possible contact needs to be detected in all three dimensions.

3 Contact Detection with the TouchDown-Sensor

The TouchDown sensor is a critical part in the alignment sequence, as the mechanical contact between tool and part has to be detected by measuring the contact force, as the SEM lacks the third dimension. Generally, three configurations are suitable for the implementation of such a sensor. At first, the senor can be integrated into the tool. As the handling tools for nano-scale assembly are fabricated in silicon batch processes, such integration is costly and the flexibility is low. Secondly, the tool can be mounted on the sensor, or as a third solution, the adaptor carrying the parts can be mounted on the sensor. The latter two approaches allow for highly flexible configurations, which are quasi independent of the tools or parts used for the assembly process. On the downside, a high sensitivity is necessary for such an approach. The initial Touch-Down-sensor is basically a piezo-bimorph cantilever: two piezoelectric layers are connected by a glass fibre substrate (Fig. 3). The upper one is used as actuator, whereas the lower one is used as receiver. This tactile sensor has been implemented as a vibrating sensor, oscillating at its resonance frequency with small amplitude [11]. The occurring contact between tool and part changes the oscillation characteristics, modelled as an additional damping element. The change in phase shift between the driving and the driven oscillation, which is measured with a lock-in amplifier, is strongly dependent on the force exerted on the sensor. A more-dimensional approach under development detects contact with the same principle. But it is designed to superimpose this sensing principle on a second movement (e.g. a circular movement) with lower frequency but higher amplitude.

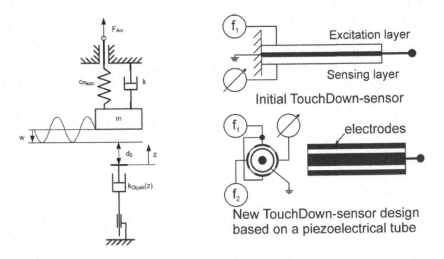

Fig. 3. Mechanical model for the TouchDown-sensor, modeled as a spring-mass-damper system. The contact between the object, e.g. CNT, is modeled as a second damper (left). The mechanical setup for both approaches is schematically depicted (right).

4 Results

The automated alignment sequence consisted of three sub-sequences. The first step (Fig. 4a) consisted of the course x,y-alignment at a scan field size of $35\mu m$. The z-course alignment was performed by depth-from-focus (cp. section 2). The second step was the fine alignment step in x-direction (Fig. 4b). The x alignment has been separated from the y alignment in order to get more reliable results from visual servoing by avoiding overlap. Finally, the metal tip moved below the CNT (Fig. 4c). The steps 2 and 3 could be performed safely because of a distance of more than $25\mu m$ between substrate and CNT (cp. Fig. 2 h1 and h2). Starting from the x-y aligned setup the touchdown sequence has been executed.

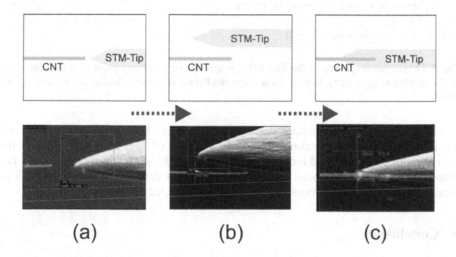

Fig. 4. Alignment sequence between a fine etched metal tip and a CNT. The top row shows the schematic drawings, the bottom row the corresponding images taken in the SEM.

Fig. 5 (left) shows the measured oscillation amplitude at resonance frequency in *nm* over the amplitude of the driving voltage in *mV*. Obviously, a wide range of amplitudes can be set with relatively small voltages. It is important that the oscillation amplitude of the sensor is smaller than the critical size of the object (CNT). This allows for even smaller CNTs or other nano-scale objects to be detected with the TouchDown-sensor.

Figure 5 (right) shows the contact detection between a fine etched metal tip and a CNT. The contact is detected as soon as a predefined relative phase shift threshold has been reached. The sensor was excited with a voltage of $0.7mV$. This corresponds to oscillation amplitude of $98nm$. The result indicates that the contact can be safely detected. The measured data showed quasi linear behaviour after the first contact with a factor of $7.135 \cdot 10^3 s/m$. The force applied to the CNT at detection threshold was about $4.9\mu N$.

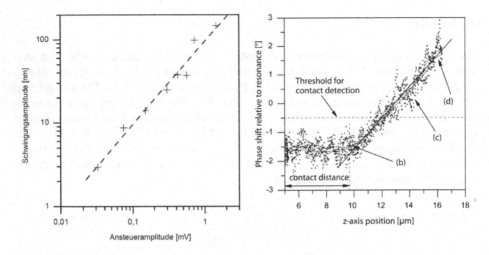

Fig. 5. Oscillation amplitude of the TouchDown-sensor in its working point over the amplitude of the driving voltage (left). Contact measurements between a fine etched metal tip and a CNT (right).

The reliability of the contact detection method has been evaluated during 500 touchdown cycles. The metal-tip started at a distance of $4\mu m$ below the CNT at the beginning of each cycle and moved at a velocity of $0.133\mu m/s$ towards the CNT. The phase shift threshold has been set to $1.5°$ for contact detection. During the 8.3h of measurement 98.4% of all touchdowns have been detected correctly.

5   Conclusion

The 3D-alignment procedure in an assembly process on the nanoscale in SEM consists of in- and out-of plane alignment steps between tool and part. Both steps are crucial for automated assembly on the nano-scale and compared to the macro- and meso-scale, dedicated tools and methods have to be developed. The TouchDown-sensor provides the necessary contact detection. Its reliability was proven in this presentation. Three-dimensional measurement would add extra value to a possible implementation of automated assembly systems. This will be achieved in further research by superimposing the TouchDown-principle on an advanced structure which is capable of more dimensional movement.

References

[1] Automatic nanohandling station inside a scanning electron microscope. 222(1), 117–128 (2008)
[2] Cecil, J., Powell, D., Vasquez, D.: Assembly and manipulation of micro devices-a state of the art survey. Robot. Comput. -Integr. Manuf. 23(5), 580–588 (2007)

[3] Eichhorn, V., Fatikow, S., Wich, T., Dahmen, C., Sievers, T., Andersen, K., Carlson, K., Bøggild, P.: Depth-detection methods for microgripper based cnt manipulation in a scanning electron microscope. Journal of Micro - Nano Mechatronics

[4] Fatikow, S., Seyfried, J., Fahlbusch, S., Buerkle, A., Schmoeckel, F.: A Flexible Microrobot-Based Microassembly Station. Journal of Intelligent and Robotic Systems 27(1-2), 135–169 (2000)

[5] Fearing, R.S., The, Kg.: Survey of sticking effects for micro parts handling. In: IEEE/RSJ Int. Workshop on Intelligent Robots & Systems (IROS), pp. 212–217 (1995)

[6] Sievers, T.: Global sensor feedback for automatic nanohandling inside a scanning electron microscope. In: Proceedings of IPROMS NoE Virtual International Conference on Intelligent Production Machines and Systems, pp. 289–294 (2006); Received the Best Presentation Award

[7] Teo, K.B.K., Chhowalla, M., Amaratunga, G.A.J., Milne, W.I., Hasko, D.G., Pirio, G., Legagneux, P., Wyczisk, F., Pribat, D.: Uniform patterned growth of carbon nanotubes without surface carbon. Applied Physics Letters 79, 1534 (2001)

[8] Dahmen, C., Luttermann, T., Frick, O., Naroska, M., Fatikow, S., Wich, T., Stolle, C.: Zunami: Automated assembly processes on the nanoscale. In: 4M/ICOMM 2009 - The Global Conference on Micro Manufacture, pp. 81–85 (2009)

[9] Vikramaditya, B., Nelson, B.J.: Visually guided microassembly using optical microscopes and active vision techniques. In: Proceeding of the 1997 IEEE International Conference on Robotics and Automation (1997)

[10] Wich, T., Hülsen, H.: Robot-based Automated Nanohandling. In: Automated Nanohandling by Microrobots. Springer, Heidelberg (2008)

[11] Wich, T.: Tools and methods for the automation of serial nano-assembly processes in the Scanning Electron Microscope. (Werkzeuge und Methoden zur Automatisierung der seriellen Nanomontage im Rasterelektronenmikroskop). Logos Verlag Berlin GmbH (2008)

Guidelines for Implementing Augmented Reality Procedures in Assisting Assembly Operations

Viviana Chimienti[1], Salvatore Iliano[1], Michele Dassisti[2],
Gino Dini[1], and Franco Failli[1]

[1] Dipartimento di Ingegneria Meccanica, Nucleare e della Produzione, Università di Pisa,
Via Bonanno Pisano 25/b - 56126 Pisa, Italy
[2] Dipartimento di Ingegneria Meccanica e Gestionale, Politecnico di Bari
Viale Japigia 182 - 70126 Bari, Italy
{viviana.chimienti,salvatore.iliano,dini,f.failli}@ing.unipi.it,
m.dassisti@poliba.it

Abstract. The use of Augmented Reality (AR) in training or assisting operators during an assembly task can be considered an innovative and efficient method in terms of time saving, error reduction, and accuracy improvement. Nevertheless, the implementation of an AR-based application is quite difficult, requiring to take into account several factors. This paper provides a general procedure to follow for a correct implementation, starting from an assessment of the assembly task, until the practical implementation. To assess the procedure, it has been applied to the training of unskilled operators during the assembly of a planetary gearbox, with the help of a hand-held device.

Keywords: Assembly, Augmented Reality, Training.

1 Introduction

Training and assisting operators in assembly procedures represents an important point in different industrial situations: precision assembly, use of temporary personnel in assembly lines, etc. Up to now, two different training methods have been adopted: on-the-job and face-to-face training. The former is accomplished directly on assembly stations and requests the continuous attendance of an experienced operator placed side by side to the unskilled one. The latter consists in face-to-face lessons given in different modalities (traditional or using computer-assisted procedures). These methods are undoubtedly quite effective, but serial, in that they consume productive time of skilled operators. A new training method is represented by the application of AR techniques, overcoming the above mentioned cons in assisting unskilled operators all over the assembly procedure, providing step-by-step instructions and thus assuring an immediate capability to accomplish the task by himself.

2 Augmented Reality in Assembly

Augmented Reality (AR) is a concept developed in the last decades, consisting in improving information content in a real environment. The basic idea, used also in

S. Ratchev (Ed.): IPAS 2010, IFIP AICT 315, pp. 174–179, 2010.

non-industrial scenarios, is to supplement a real scene by synthetic images superimposed on it. The goal of AR is therefore to generate images somewhere on the optical path between the eyes of the operator and real objects in the working area.

Among the techniques used to achieve the augmentation, Video Mixing enables the user to watch the real scene indirectly through a video camera; a computer acquires the information and includes the digital content. Real and virtual objects coexist as two separate video streams, and the result is shown on a display. In [1] other techniques (i.e. Optical Combination and Image Projection) are described. The hardware used is[1][2]: a camera, to frame the real environment; a computer, that creates virtual contents and mixes real and virtual video streams together; a display, which shows the results of the augmentation; a tracking system, to detect operator's mutual position with the camera. A taxonomy of the different displays used can be found in [3]; considering their position compared with operator, the following can be distinguished: Head Mounted Display (HMD), Hand Held Display (HHD) and Spatial Display (SD).

Several applications have been investigated in the field of assembly so far (e.g. [4][5]). A first example was applied in aircraft industry, to electrical wiring assembly: the path of the wire is shown to the operator on a HMD in order to follow the visual track to perform the wiring operations. In automotive industry, AR methods were used in door lock assembly, with the target of creating a training instrument.

One of the main problems in this kind of applications is represented by the complexity of the AR implementation procedure: an effective application of this innovative technique requests the analysis of several aspects related to the assembly procedure. The aim of this work is to propose standard guidelines for a correct implementation of AR systems for guiding operators while assembling products, obtaining the advantages of time saving, error reduction and accuracy improvement.

3   Proposal of Implementation Procedure

The implementation procedure proposed in this paper is illustrated in Fig.1. The goal is to create a "standard procedure" to be followed whenever an AR method has to be applied for supporting an operator in performing an assembly task. The procedure is described in the following paragraphs using, as test case, a planetary gearbox (Fig.2) consisting of over than 200 different parts. Despite the specific application, the procedure is general enough to be adopted for other assembly operations.

Preliminary Analysis of the Assembly Procedure. The process starts with the analysis of the product, its parts, the assembly sequence, presence of subassemblies and the identification of assembly relationships among components. The purpose of the analysis is to have a clear vision of the components and of the process under study, checking all the elements to be manipulated and the information the operator might need in addition. During this step, the list of components related to the gearbox has been created; 4 assembly groups (output, 2nd stage, 1st stage, input) and 2 assembly subgroups (1st stage and 2nd stage planetary gear) were identified.

Subdivision in Tasks, Sub-Tasks and Elementary Operations. The assembly process is hierarchically divided in tasks, sub-tasks and elementary operations. Going down in level of detail, the operation becomes more and more elementary and indivisible in other sub-operations. Each action will be then described accurately and all relevant information to perform correctly the operation will be identified, such as tools, devices, equipment, safety requirements, and organized in a table, as shown in Table 1, referring to the assembly of the output group of the gearbox.

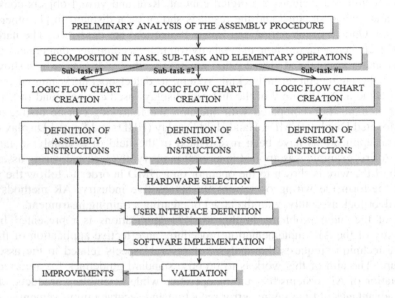

Fig. 1. Standard procedure for implementing an AR application for assisting operators during assembly tasks

Fig. 2. Case study: planetary gearbox

Creation of Logic Flow-Charts. Each task and sub-task should be represented by means of logic flow-charts, including the assembly sequence, check points, variants and alternative procedures. These charts will be used to carry out the software implementation. Compared to the definition of the assembly cycle described above, this diagram allows to go through alternative paths, if there are checkpoints. Fig.3 reports an example of the flow chart regarding task #2, namely the assembly of the roller bearing on the shaft.

Definition of Assembly Instructions. For each elementary assembly operation, suitable instructions have to be identified. This step includes the selection of textual messages, icons, 2D pictures or 3D models which have to be positioned in the real environment through AR. An example is reported in Table 2.

Table 1. Tasks, Sub-Tasks and Elementary Operations in assembly of the output group

Task	#	Sub-task	#	Elementary operation	Tools	Safety requirements	Remarks
Assembling ball bearing on the casing	1	Press equipping	1	Take output casing		Wear gloves	
			2	Place it on the base	Base "K"		
			3	Lubricate bearing housing	Oil		
				.. [...]			
	2	Pressure on the bearing	1	Pump oil to bring piston near the punch			No contact!
			2	Ensure the centering			Visually
			3	Close protection cover		Close cover!	
				.. [...]			
	3	Coupling check	1	Check if the feeler gauge goes through bearing and housing	0.05mm Feeler gauge		
				.. [...]			
Assemblying output shaft-roller bearing	1	Press equipping	1	Take output shaft		Wear gloves	
				.. [...]			

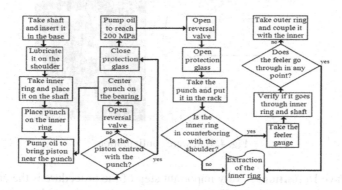

Fig. 3. Flow chart of the assembly of the roller bearing on the shaft

Table 2. Correspondence between assembly operation and virtual elements to be visualized in AR environment

Elementary operation	Text message	Symbol	Image	3D model	Other images
SHAFT LUBRIFICATION	Oil shaft where arrow points				
INNER RING POSITIONING	Set the inner ring on the shaft				

Hardware selection. The AR hardware is selected according to the main features of the working environment and to the assembly process to be performed. Selection charts as the one illustrated in Table 3 are used to choose the most appropriate device.

Table 3. Example of a selection chart used for choosing the most appropriate hardware device

Hand Held Display		Head Mounted Display		Spatial Display	
Pros	**Cons**	**Pros**	**Cons**	**Pros**	**Cons**
Integrate in one device camera, display and processor	Low performance of processors used	Good integration between real and virtual	Low confort	No ergonomic problems	Occlusion of the projection by objects or the user
Easy to find or purchase	One hand is not free	Portable	Fixed image depth	No visual fatigue	Only for fixed applications
Non invasive				Wide displays	

For the gearbox assembly, a handheld device has been chosen, consisting of an 8" touch screen monitor (Fig. 4). The tracking system selected was the optical marker-based tracking, with the software for the management of real and virtual streams.

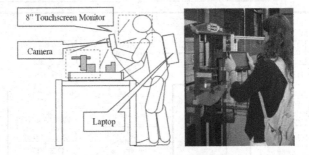

Fig. 4. Hardware configuration

User Interface Definition. A very important step of the procedure is the creation of a graphical user interface (Fig.5). The system functionalities should be easily perceived by all kinds of users. So, it is a good choice to put more emphasis on clarity and abundance of information available, distinguishing between essential information for any type of operator, the ones that immediately appear on the screen, and information that are shown only if operations are carried out by less experienced staff.

Software Implementation. The planned procedure is implemented by programming the AR software and by preparing the working environment with AR tools and other devices (tracking system, hardware docking stations, etc). The lightening of the work environment needs to be considered accurately, since it influences software's markers recognition: too much or insufficient lightening can cause problems. Depending on the chosen software, different ways of implementing can be adopted, such as writing a C# code, or using software's already implemented *actions*, also written in C#, which have to be recalled graphically to create the workflow.

Validation. The AR system should be validated using a sample of users having different levels of experience and competences. A questionnaire can be proposed for collecting responses, comments and difficulties encountered in performing the assembly tasks, to be used for enhancing the implementation made during the previous step. Questions must be formulated appropriately in order to analyze procedural fairness, clarity of instructions, ergonomics, effectiveness and efficiency of AR in assembly training.

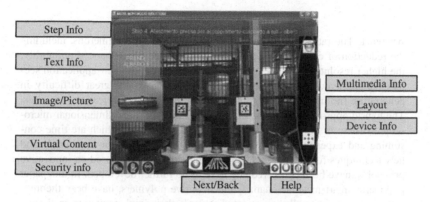

Fig. 5. Graphical User Interface definition

4 Conclusions

The proposed guidelines offer an effective starting point for the implementation of an AR training system for assembly operations, and their general nature gives it the flexibility to make it applicable also in other fields. Compared to the training techniques used previously, the AR application seems to offer a set of advantages, such as a significant reduction of time and a lower investment in human resources.

Concerning the validation test performed on the gearbox, the specific solutions implemented was also successful thanks to the application of the procedure. Technical features selected were positively appreciated by operators. The whole sample of people tested was able to successfully conclude the procedure in timing between 20 and 35 minutes.

References

1. Azuma, R.T.: A Survey of Augmented Reality. Presence: Teleoperators and Virtual Environments 6(4), 355–385 (1997)
2. Azuma, R., Baillot, Y., Behringer, R., Feiner, S., Julier, S., MacIntyre, B.: Recent Advances in Augmented Reality. IEEE Computer Graphics and Applications, 34–47 (November/December 2001)
3. Milgram, P., Kishino, F.: A taxonomy of mixed reality visual displays. IEICE Transactions on Information and Systems E77-D(12), 1321–1329 (1994)
4. Pang, Y., Nee, A.Y.C., Ong, S.K., Yuan, M.L.: Assembly Feature Design in an Augmented Reality Environment. Assembly Automation Journal 26(1), 34–43 (2006)
5. Pang, Y., Nee, A.Y.C., Youcef-Toumi, K., Ong, S.K., Yuan, M.L.: Assembly Design and Evaluation in an Augmented Reality Environment. In: Singapore-MIT Alliance Symposium, Singapore, January 19-20 (2005)

Monodirectional Positioning Using Dielectric Elastomers

C. Pagano, M. Malosio, and I. Fassi

Institute of Industrial Technology and Automation, CNR Via Bassini 15, Milan, 20133, Italy

Abstract. The rationales for the use of microsystems are numerous, including the reduction of consumables, a faster response time, the enhanced portability, the higher resolution, and the higher efficiency; moreover their application sectors are numerous. Nevertheless microproducts have still great difficulty in penetrating the market, mainly due to the limits of the fabrication processes. The hybrid approach is suitable for the fabrication of three dimensional microscopic structures but often requires manual contributions, which are time consuming and expensive. In order to overcome these issues, new materials and new techniques for the manipulation of microcomponents, based on innovative principles, have been conceived and have to be further developed. In this paper polymeric smart materials, namely electroactive polymers, have been theoretically and experimentally investigated towards their implementation in the actuation and sensing of positioning and handling devices. The feasibility of a monodirectional positioner has been studied.

Keywords: Electroactive polymers, microfactory, microactuator, hybrid microproducts.

1 Introduction

Despite the advanced state of conventional actuation technologies, there is an increasing demand for new actuation devices with high power-to-weight ratio, high efficiency and large degree of compliance. In order to fulfill these needs, innovative types of actuators based on smart materials have recently been proposed. They are based on polymeric materials able to change their dimension and/or shape in response to a specific external stimulus such as thermal, chemical, electrical, magnetic or optical. A promising group of such materials is represented by the Electro Active Polymers (EAPs), which exhibit interesting properties, including a controllable high ratio strain versus applied voltage, low specific gravity, high grade of processability, downscalability and, in most cases, low costs [1]. Nevertheless, successful implementation of electroactive polymers mainly depends on the degree of understanding of their behaviour and properties since the selection of the material in the design of smart systems involves considerations of different factors, including the maximum achievable strain, stiffness, spatial resolution, frequency bandwidth and temperature sensitivity. Moreover, it is still essential to develop innovative testing methodologies, consistent material characterization and formulation of advanced theoretical models [2].

In the next section a short introduction on electro-active polymers is presented with particular attention on dielectric elastomers. In the third section the material and the

S. Ratchev (Ed.): IPAS 2010, IFIP AICT 315, pp. 180–187, 2010.

realization process of the samples are illustrated. In the fourth section the experimental analyses are described and the results discussed. In the fifth section the concept of an EAP-based planar positioner is presented and, finally, the conclusions are drawn.

2 Electroactive Polymers

Electroactive Polymers are polymeric materials whose shape can be modified by applying a voltage to them. The large displacement that can be obtained with EAPs, their low mass, low power and, in some of these materials (ionic polymer) also low voltage, make them very attractive as actuators. Moreover, their mechanical flexibility and ease of processing offer advantages over traditional electroactive materials expanding the options for mechanical configurations. The group of EAPs comprises a wide array of different materials, each characterized by its unique properties and functional abilities [2].

Dielectric elastomers are among the most promising EAPs for many applications including actuators and sensors for the microfactory: they work in a dry environment, can achieve great deformations and support high voltage. When an electric potential difference is applied across the polymeric film, coated with electrodes on both sides, a compressive stress, parallel to the electric field, is generated. This is a well-known phenomenon, known as Maxwell stress, which occurs with all insulators subject to an electric field from deposited electrodes. Nevertheless, only the recent development of soft polymeric thin films with high dielectric breakdown strengths allows the achievement of performances exploitable in actuation devices [3]. EAP operation principle is similar to the electromechanical transduction of a parallel two plate capacitor. When an electric potential difference is applied across the polymeric film, coated with electrodes on both sides, a compressive stress, parallel to the electric field, is generated and, consequently, the material is compressed in thickness and expands in planar directions. In order to permit this deformation the electrodes in an EAP actuator must be compliant because the elastomeric polymers are essentially incompressible in volume and so when the polymeric film is compressed in thickness, it must expand in area [4]. Mechanical actuation can be obtained by the compressive stress which causes a compression along the thickness of the material and the consequent planar expansion of the dielectric elastomer. The electrostatic pressure that compress the film along the thickness direction has been derived via the principle of virtual work assuming that the electrodes are much more compliant than the polymer film itself [5]. This effective pressure generated on the film can be expressed by the following equation:

$$P = \varepsilon \, \varepsilon_0 \, E^2 = \varepsilon \, \varepsilon_0 \, (V/t)^2 \tag{1}$$

where ε_0 and ε respectively denote the permittivity of the free space and the dielectric constant of the polymer; E and V denote the applied electric field and voltage, and t is the thickness of the film. Because of the compliance of the electrodes, P results to be twice higher than the pressure in a rigid parallel plate capacitor [5].

3 Materials and Methods

EAPs are relatively new materials and, even though their potential applications are spread out in many different sectors, very few EAPs are commercially available and no exhaustive characterization on their electromechanical behaviour exists. For this reason, in order to evaluate their actual limits and potentialities, some experiments have been carried out with a commercial EAP (the 3M acrylic elastomer VHBTM 4910). It is one of the most studied EAP and its electrostrictive behaviour is by now widely known [6]-[10]. The material is a clear and colourless film, with a dielectric constant of 4.7 in the un-stretched state. It is very sticky and can be largely stretched in both directions in the plane of the film. In our experimental setup, firstly, the polymer samples have been pre-stretched up to 700% along each linear dimension through a circular and adaptable stretching device. Secondly, the film has been packed between a thick and a thin rigid support (20mm x 26mm), both of which present a window of 11mm x 20mm in the centre, in order to obtain a pre-stretched completely constrained sample. In the bare and central region the film is free to deform in consequence of an external electric field. Thirdly, a region with the shape of a stripe (20mm x 2mm) on both sides of the film, inside the window and adjacent to one of the bound regions, has been coated with a conductive material. These main components are illustrated in Figure 1. Finally, some markers have been imprinted to one side of the film, through an inked grid-stamp. A sketch and a picture of a sample, together with the reference axes, are shown in Figure 1Figure 2 and Figure 3 respectively. In order to obtain appreciable displacements, the samples have been pre-stretched so that the electric field can be increased without reaching the breakdown voltage [11], in accordance with equation (1).

The shape and the dimension of the electrodes respect to the mechanical constraints have been chosen in order to obtain an almost monodirectional in-plane deformation of the film, along y axis, whereas the deformation in x direction can be neglected. This allows the direct transposition of the decrease of the thickness to the deformation along only one direction.

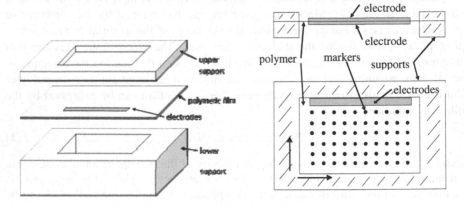

Fig. 1. Exploded view of the experimental sample

Fig. 2. Sketch of the sample: section view and top view

The choice of the material for the electrodes is an issue concerning the fabrication of EAP actuators. Indeed, not only have they to be robust enough to endure high electric fields [12], up to hundreds of kV/mm, but they have also to be very compliant and maintain a good electrical contact when stretched. Graphite powder and silver grease have been tested, in order to obtain highly conductive electrodes. The grease has showed better conductance; hence, Chemtronics Circuit Works™ silver conductive grease has been preferred and, in some samples, graphite powder has been added to increase the conductivity. Through a suitable vision system the displacement of the markers has been measured in order to quantify the deformation of the polymer when an electric field is applied.

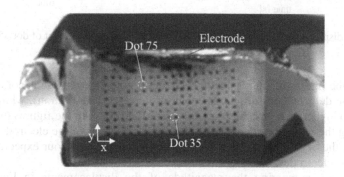

Fig. 3. Picture of a sample with the system reference

Several samples have been prepared and analyzed applying different values of the external voltage, from 1kV to 6 kV with steps of 1kV. The voltage has been applied to the conductive region of the polymer, coated with the compliant electrode, through a small and thin aluminium wire and the reaction of the central region of the polymer has been recorded through a vision system, equipped with a telecentric objective. The collected images have been elaborated via image processing techniques in order to obtain information about the position and the displacement of the markers and, therefore, the planar deformation of the polymer. The charge/discharge cycles procedure is here described: firstly the lowest voltage (1kV) has been applied and turned off for few seconds and then, turned on again for 5-7 times. The same procedure has been followed applying 2, 3, 4, 5 and finally 6kV and between two following charge/discharge cycles a time interval of approximately 2 minutes have been waited to identify possible residual deformations, permit to overcome transitories and avoid failures due to significant loads for long periods of time [13]. All the process has been recorded by the vision system with a 30 frames-per-second rate. Moreover all the charge/discharge cycles have been analysed at once, tracing the displacements of the dots with respect to their positions in the very first image (before applying 1kV) and over the entire period of time. The displacement along x and y of a dot near the electrodes (dot 75 in Figure 3) and one far away from the electrodes (dot 35 in Figure 3) has been plotted versus the time in Figure 4 and Figure 5 respectively.

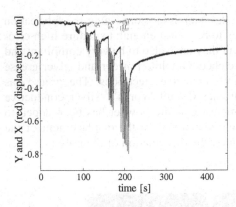

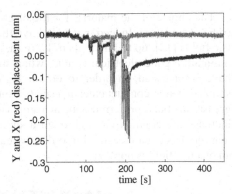

Fig. 4. X-Y displacement of dot 75 in Figure 3 vs. time

Fig. 5. X-Y displacement of dot 35 in Figure 3 vs. time

In both the figures the deformation cycle due to the 1kV tension is not appreciable (because the deformations are comparable to the camera resolution), while deformations due to higher tensions can be clearly identified. In both the figures the displacements along the y direction (perpendicular to the main axis of the electrode) are much bigger than the ones in the x direction; this confirms the behaviour expected from the layout of the electrodes.

Furthermore, comparing the magnitude of the displacements in Figure 4 and Figure 5 it is evident that, as expected, the closer to the electrodes the dot, the bigger the displacement. Finally, a residual deformation is evident in both the plots, and, as expected, a greater residual deformation is associated with the dot nearer to the electrodes (dot 75 in Figure 4).

In Figure 6 small movements are notable along x almost only in the upper part of the plot where the movement along the y direction are significantly small and effects of mechanical constraints can much more influence the deformation of the polymer. Further analyses of the displacements along x will be carried out in order to estimate the influence of boundary effects.

In Figure 7 the maximum measured displacement in the y direction of two samples with different thickness (28 μm and 62 μm) has been plotted versus the applied voltage.

The thickness of the samples has not been experimentally measured, but evaluated from the magnitude of the mechanical pre-stretch. Indeed, since the Poisson ratio of the material is equal to 0.49 [14], it has been considered incompressible and the thickness of the sample has been calculated according to the following formula:

$$x^2 + y^2 + z^2 = \text{constant} \tag{2}$$

where x, y and z are the three dimensions of the film.

The displacement increases in a non linear way when the applied voltage is increased and the results obtained for both the samples have been interpolated with a second order polynomial curves, shown in Figure 7. The consistency of the interpolating curves with the results seems to be in agreement with equation (1) even if

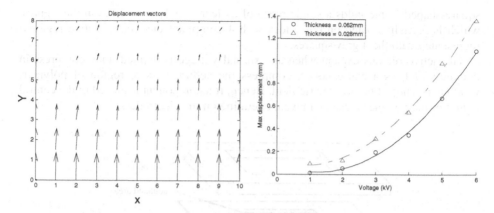

Fig. 6. The arrows refer to the displacement field of the sample, their direction and module represent the maximum displacement of the reference points on the sample

Fig. 7. Absolute values of the maximum displacement along the y direction versus the applied voltage

further investigations on the deformation along the z direction have to be carried out to fully understand the behaviour of the materials. Moreover, this behaviour is in agreement with the hypothesis according to which the electrical compression in the z direction (perpendicular to the plane of the polymeric film) is mainly transformed in a stretch along the y, whereas the x displacement is strongly limited by the bounds and the asymmetric shape of the electrode region. Finally, it can be noticed that the two curves are deformed along the y coordinate due to the different thickness of the film of the samples. Indeed, the voltage applied to both the samples is the same (from 1kV to 6kV), but, due to the thickness, the electric field strength is different, according with equation (1). Finally, Figure 7 shows that displacements bigger than 1mm can be obtained with 6kV and that a range of about one order of magnitude can be achieved varying the voltage between 1 and 6kV.

4 Planar Positioner

Electroactive polymers are very promising in a variety of sectors and, in particular for applications in the Microfactory. Indeed, due to the light load and the reduced space available, the devices required at small scale have specific needs, such as lightness, compactness, precision and rapidity, which are essential requirements for microactuators together with common needs, such as suitable force and stroke.

The idea behind the actuated samples can be directly converted into a monodimensional linear actuator. In order to realise a position-controlled device two opposite couples of electrodes are required, so that the actuation of each of them can recover residual deformations and histeretical behaviour of the material, due to a previous activation of the opposite one.

Moreover, a more advanced bi-dimensional micropositioning actuator, made of dielectric elastomer has been conceived. In Figure 8 a sketch of it is drawn: a rigid square-shaped frame holds a single piece of dielectric polymer (white in the figure), which is partially coated on both sides with 4 compliant electrodes (active regions) represented with the 4 grey squares.

The active regions expand when an external voltage is applied, they compress in thickness and, as a consequence, compress the central passive region of polymer, which is no subject to any electric field. Hence, if a component is placed in the central region of the polymer it can be moved, operating on the electrodes.

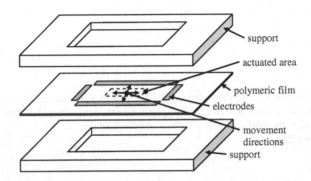

Fig. 8. Sketch of a potential configuration of a micropositioning made of dielectric elastomer

This configuration allows the activation of one or more active regions at time and also the recovery of residual deformations and the precise control of the position of the central region in the plane in two dimensions for planar positioning applications.

5 Conclusions

Although a lot of potential applications of electroactive polymers could enable many innovations in several fields, their diffusion is still limited by the lack of clear understanding of all their properties. Indeed their structure-property response interrelations is complex and a further issue is due to the sensitivity of electroactive polymers to their fabrication conditions, time and temperature dependent effects, and nonlinearity of their coupled mechanical, physical and chemical properties.

In this paper the electro-mechanical behaviour of an acrylic EAP has been studied and a few physical properties, such as the strain for different input loads and the presence of a residual deformation have been observed. Moreover an application of the polymer in the microfabrication of hybrid MEMS has been introduced, taking into account the results obtained. Further analyses will be carried out in order to characterize and model the hyperelastic behaviour of the material both under a theoretical and an experimental point of view. On the basis of the experimental results, a model-based control system will be studied, in order to realise a positioning device with submicrometric precision as required by the application sector.

References

[1] De Rossi, D., et al.: NTS Inc., pp. 119–135 (2004)
[2] Vinogradov, A., et al.: Mater. Res. Soc. Symp. Proc., 889 (2006)
[3] Bar-Cohen, Y.: SPIE Press, Bellingham (2004)
[4] Pelrine, R.E., et al.: Adv. Mater. 12/16 (2000)
[5] Pelrine, R.E.: Sensor Actuator A 64, 77–85 (1998)
[6] Jung, K., et al.: Smart Mater. Struct. 16, S288–S294 (2007)
[7] Kofod, G., et al.: Proc. of SPIE, vol. 4329, pp. 141–147 (2001)
[8] Pelrine, R.E., et al.: Science 287, 836 (2000)
[9] Wissler, M., Mazza, E.: Sensors and Actuators A 134, 494–504 (2007)
[10] Carpi, F., et al.: Sensors and Actuators A 107, 85–95 (2003)
[11] Kofod, G., et al.: J. of Intelligent Material Systems and Structures 14(12), 787–793 (2003)
[12] Santer, M.J., Pellegrino, S.: Proc. of the 45th AIAA/ASME/ASCE/AHS/ASC Structures, Palm Springs, California, April 19-22 (2004)
[13] Plante, J.S., Dubowsky, S.: Smart Mater. Struct. 16, S227–S236 (2007)
[14] 3M VHBTM Tapes Technical

Chapter 7

Gripping and Handling Solutions

Active Gripper for Hot Melt Joining of Micro Components

Sven Rathmann, Annika Raatz, and Jürgen Hesselbach

Institute of Machine Tools and Production Engineering,
Technical University Braunschweig,
Langer Kamp 19b, 38106 Braunschweig, Germany
s.rathmann@tu-bs.de

Abstract. Precision assembly of hybrid micro systems requires not only a high precision handling and adjusting of the parts but also a highly accurate and fast bonding technique. In this field adhesive technology is one of the major joining techniques. At the Collaboration Research Center 516, a batch process based on a joining technique using hot melt adhesives was developed. This technique allows the coating of micro components with hot melt in a batch. The coating process is followed by the joining process. Due to this, the time between coating and joining can be designed variably. Because of the short set times of hot melt adhesives short joining times are possible. For this assembly process adapted heat management and adapted gripper systems are very important. Primarily, the conception and construction of suitable grippers, which realize the complex requirements of the heat management, meet a high challenge. This paper presents an active gripper system for the joining of micro parts using hot melt adhesives. Furthermore, first examinations and results of an assembly process using this gripper for bonding with hot melt adhesives are presented.

Keywords: Precision joining operations, hot melt adhesives, active gripper systems.

1 Introduction

The necessity of integrating more and more functions into micro components by unchanging or decreasing components' size makes a high precision assembly in the sub micron area more and more important [1]. To reach this goal, many endeavors in the field of high accurate handling and positioning systems were and are made [2, 3, 4]. Another very important part is the use of a suitable bonding technology. The most important technology is the adhesive bonding technique [5]. Requirements to this bonding technique are: a rapid fast curing time, the possibility of a pre-application of the adhesive, a low or nonexistent shrink of the adhesive and the possibility to use this bonding technique in batch processes. Nowadays, chemical or light reacting adhesives are used. These adhesives normally achieve high strength of the bond. But the disadvantages of these adhesives are their long curing times (light reacting adhesives have short curing times, less than 10 seconds, but it is not always possible to irradiate the bonding area), the constraint to apply the adhesive directly or shortly before the joining process and their low suitability for batch processes. The approach developed in

S. Ratchev (Ed.): IPAS 2010, IFIP AICT 315, pp. 191–198, 2010.
© IFIP International Federation for Information Processing 2010

the Collaborative Research Center 516 "Design and Manufacturing of Active Micro-systems" is the use of hot melt adhesives. The main advantages are extremely short set cycles, the possibility of pre-applying the adhesive and the time-delayed joining procedure [6]. A disadvantage is their low adhesion. The comparison of the advantages and disadvantages shows that this bonding technology is an interesting alternative for the assembly of hybrid micro systems.

To establish this technology, adhesive and application technologies as well as the assembly process have to be developed. Furthermore, the process parameters must be determined. A very important aspect for the process design is the kind of heating technology that should be used. Therefore, special process components, such as fixtures and grippers as well as the heat management concepts should be developed. In [7] two approaches to solve these problems are described. The following section discusses these concepts and augments the approaches using a phase model. The assembly process will also be described. In the third section a new active gripper concept will be presented. It can be used for joining micro components with hot melt adhesives. Finally, first results of test assemblies will be shown and discussed.

2 Adapted Assembly Strategies for Hot Melt Joining

Heat management is an integral part of the selection and the design of an assembly process using hot melt adhesives, as the volume of the hot melt is quite small and the thermal capacity is rather low. This results in the hot melt not being able to store much thermal energy. Thus heating the hot melt adhesive directly is not recommendable. On the contrary the thermal capacity of the component and the gripping system is much higher. Hence the characteristics of the grippers and the components such as thermal capacity and thermal conductivity are much more influential to the heat management. To achieve this, special considerations have to be taken in account when designing a specific assembly process, as the choice of a suitable heating source is crucial for process design. For heating up the gripper unit there are multiple devices available: heating plates, infrared heaters, lasers, heating foils, heating ceramics and Peltier elements, or a combination of these.

In general, the concepts of heat management can be split into a passive and an active concept [8]. The passive heat management concept makes use of the principle of heat storage to supply the energy for the joining process. The active concept carries a heat source inside the gripper system. Depending on the design, the heat emission has to be controlled in order to customize the heating to specific needs of the process.

The solid line in Figure 1 indicates a temperature profile of the passive heat management concept. The joining sequence is divided into three consecutive phases: a heating phase, a processing phase and a post-processing phase. During the first phase, the heating phase, the gripper and the component are heated by a heat source to the working temperature T_{Hp}. The handling, e.g. movement of the component from the heat source towards the target position, is carried out in the processing phase, which is limited by t_{Hp} and t_{Kp}. When the component touches the substrate the post-processing phase begins. The temperature drops substantially, the hot melt sets and reaches its final strength.

In contrast to the passive heat management concept, the heat source of the active heat management concept (dashed line in Figure 2) is integrated into the gripping or clamping unit. With this setup heat can be continuously transmitted into the component during the handling or the joining process. Thus the temperature can be controlled and kept at a constant level during the processing phase or heating can instantly be applied when necessary. The working temperature is thereby dependent on the melting temperature and not on the process time. This is why the processing temperature can be set at a lower level.

The advantage of the active heat management concept is the independence of the assembly process from the heat capacity of components which allows flexibility in the process design.

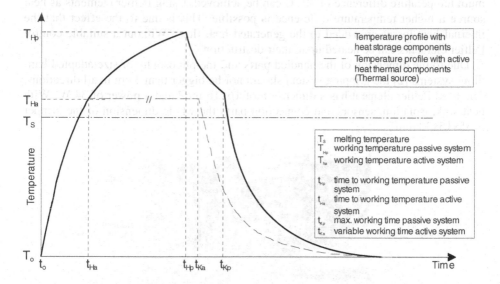

Fig. 1. Temperature profiles on different heat management concepts

Consequently, the assembly of micro components has to accomplish the following steps to follow the described heat management concept:

1. gripping the hot melt coated component
2. heating up the component (heating phase)
3. aligning the component with the joining position (processing phase)
4. joining (post-processing phase)
5. wait until the component temperature has dropped below the melting temperature (post-processing phase)
6. release the bonded component

In the following examinations and experiments the hot melt adhesive "Vestamelt 732" will be used. This adhesive has its melting temperature T_S in the area of 100 to 110 °C. Experiences with this adhesive show that a working temperature T_{Hp} of about 140 °C is suitable for joining with this adhesive.

3   Active Gripper System

Figure 2 shows the active gripper concept. It consists of a heat source, which is carried out as Peltier element, a heat sink, a vacuum gripper and an isolation to the robot system. The bottom side of the gripper will be called primary side. This side is important for the heat exchange to the component. The side of the Peltier element which is connected to the heat sink will be called secondary side.

Peltier elements are heat pumps. Depending on the impressed current direction the Peltier element generates a heat flux from one to the other side of the element. Thus a temperature difference between the sides will be generated. Normally, Peltier elements will be used as thermoelectric cooler. With standard Peltier elements a maximum temperature difference of 70 °C can be achieved. Using Peltier elements as heat source a higher temperature difference is possible. This is due to the effect that the internal heat losses are added to the generated heat flux . Without a suitable control Peltier elements can be heated up to their destruction.

Due to the dimension of the handled parts and the philosophy of size adapted handling systems [2] the gripping system should not be bigger than 3 cm in all directions. The used Peltier element has a dimension of 16x16 mm² and a power of 34 W. With heat sink, isolation, sensors and connection parts the whole dimension of the gripper is 35x15x30 mm³.

Fig. 2. Active gripper system

In the heating mode of the active gripper the heat flux will be generated from the secondary to the primary side. Therefore, a fast heating of the gripped component and the coated adhesive will be done. The gripper is controlled by a bang-bang controller. Due to the heat inertia this simple control concept reaches a suitable quality of control of ±2.5 °C. The parameters of the controller are selected in such a way that a mean gripper temperature of about 140 °C is reached. Thus the working temperature T_{Hp} of about 130 °C on the bottom side of the component adjusts itself.

By switching the current direction to the second mode of the gripper system, the cooling mode can be used. In this mode the heat flux from the primary to the secondary side will be generated. With sufficient heat flow at the heat sink a fast cooling effect on the primary side can be achieved. Thus it is possible to reach further reductions of the joining time.

4 Experiments

4.1 Assembly Process and Parameters

For the experiments and the validation of the active gripper system a size adapted assembly system was used. Fig. 3 shows the assembly system. The robot (micabo-f2) provides four degrees of freedom, for part handling and an additional degree for focusing the integrated 3D-vision sensor. The workspace is 160 x 400 x 15 mm3 and thus enables flexible assembly of various micro systems. Therefore, two assembly fixtures of size 4, according to DIN 32561, could be prepositioned in the clamping device. The micabo-f2 has a repeatability of 0.6 micrometers with a standard deviation of 3σ as measured according to EN ISO 9283 [9]. The standard control of the assembly system was enhanced for the control of the active gripper. Thus a fully automatic assembly process is possible.

Fig. 3. Assembly robot micabo f2 (IWF Braunschweig)

To detect possible shifts of the components during the joining process a continuously detection of the components with the integrated 3D-vision system is possible. Furthermore the joining force can be measured with an integrated 3D force sensor. In Fig. 3 the test components are also shown. During the assembly process the smaller part (3x2 mm²) will be gripped and joined with the bigger part (8x4 mm²). On the bottom site of the smaller part the hot melt adhesive is pre-applied in form of dots with a diameter of ca. 500 µm and a height of ca. 100 µm. The assembly process was implemented as described in section 2.

Fig. 4 shows the temperature profile of an exemplary assembly process. The temperature curves of the primary and of the secondary side are displayed. Marked are the three phases of the heat management.

The curves show an assembly process starting with a cold gripper, whereby the duration of the heating phase is about 21 s. If the gripper is in a stabile working state, the time for heating can be reduced to about 12 s (Fig. 5). The process phase takes about 10 s. In this phase the alignment and the joining process takes place. The components will be aligned with a maximum positioning error of 1 µm. The positioning correction will be done with a lock and move algorithm. During the joining process, the assembly system adjusts the components with a gap of about 50 µm between the components. The joining force was nearly constant during the test assemblies and was in the

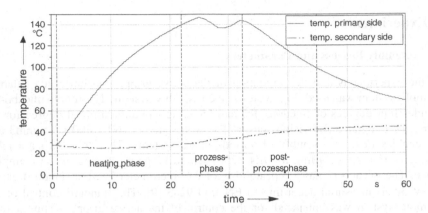

Fig. 4. Temperature profile of an assembly process

area of about 1 N. During the post-process phase the temperature of the primary side drops below 100 °C the within about 12 s and the hat melt adhesive sets. After the post-process phase the part can be released. The total assembly process takes less than 50 s which is much faster than joining processes with chemical reacting adhesives.

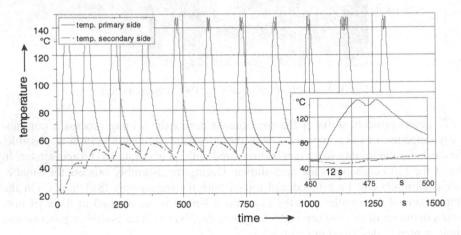

Fig. 5. Assembly cycles with stable temperature state of the secondary side

During a typical assembly processes many single assembly steps will take place. Since in these processes there are usually no long waiting times between the assembly steps, a quasi stable temperature state above the room temperature of the secondary side will be reached (Figure 5). This temperature depends on the thermal parameters of the gripping system and the parameters of the assembly process. At a process time of 10 s and a cooling temperature of the primary side of about 50 °C a process cycle time of about 120 s can be reached. In this case the temperature of the secondary side lies in the range of 45 to 55 °C.

4.2 Assembly Results and Discussion

The accuracy of assembly is an important criterion for the assessment of a newly developed assembly process. Therefore, assembly examinations with the parts described in section 4.1 were done and the uncertainty of assembly was measured. To evaluate the effect of the joining technology, the positioning uncertainty before joining and the assembly uncertainty after joining were measured. However, the obtained deviation described not only the effect of the joining processes but also the influence of the handing systems. This could be a lateral displacement of the part at the surface of the vacuum gripper and also a displacement in the magazine.

The uncertainty of assembly of the described assembly process was 11.9 µm. Less the position uncertainty of 1.4 µm the portion of the joining process and the handling components to the assembly uncertainty is 10.5 µm. Compared to the accuracy of assembly process using the passive heat management (69 µm [10]) a reduction of the uncertainty of 85 % using the active heat management could be reached.

A great part of the joining uncertainty can be ascribed to the big size variation of the adhesive dots. Furthermore, the accuracy of the positioning in z-direction depends on the variation of the component height and the measuring of the z- distance between the components resulted from the 3D-vision sensor.

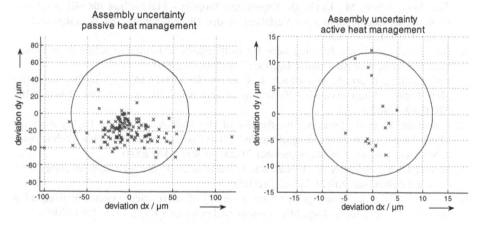

Fig. 6. Uncertainty of the assembly process (left: passive management, right: active management)

5 Conclusion and Outlook

In this paper an alternative joining process for hybrid micro systems based on hot melt adhesives was presented. Based on the outcomes of the passive heat management and associated gripper systems, a gripper system for the active heat management was presented. The heat source of the gripper based on the Peltier effect and can be used as cooling element also. The assembly examinations show the functionality of the active gripper system and offer good results for the assembly accuracy. Further increase of the accuracy can be reached by improving the holding force of the gripper, the placement accuracy in z-direction and optimizing the application process of the hot melt dots. To reduce the process time of the assembly process the active cooling effect of the gripper can be used. In this field further investigations have to be done.

Acknowledgments. The authors gratefully acknowledge the funding of the reported work by the German Research Center (Collaborative Research Center 516).

References

1. Bauer, G., et al.: Flexibles Montagesystem für die Feinwerk- und Mikrotechnik. Maschinenmarkt 105, 30–35 (1999)
2. Schöttler, K., Raatz, A., Hesselbach, J.: Precision Assembly of Active Microsystems with a Size-Adapted Assembly System. Micro-Assembly Technologies and Applications, 199–206 (2008) ISBN 978-0387-77402-2
3. Gaugel, T., Bengel, M., Malthan, D.: Bulding a Mini-Assembly System from a Technology Construction Kit. In: Proc. of International Precision Assembly Seminar (IPAS), Bad Hofgastein, Astria, March 17-19 (2003)
4. Clavel, R., et al.: High Precision Parallel Robots for Micro-Factory Applications. In: Robotic Systems for Handling and Assembly, pp. 285–296 (2005)
5. Zäh, M.F., Schilp, M., Jacob, D.: Kapsel und Tropfen - Fluidauftrag für Mikrosysteme. Evolutionäre und revolutionäre Verfahren in der Dispenstechnik. Wt-Werkstattstechnik online 92(9), 428–431 (2002)
6. Böhm, S., et al.: Micro Bonding using hot melt adhesives. Journal of Adhesion and Interface. The Society of Adhesion & Interface 7(4), 28–31 (2006)
7. Rathmann, S., Raatz, A., Hesselbach, J.: Concepts for Hybrid Micro Assembly using Hot Melt Joining. Micro-Assembly Technologies and Applications, 161–169 (2008) ISBN 978-0387-77402-2
8. Rathmann, S., et al.: Strategies for the Use of Hot Melt Adhesives for Highly Accurate Micro Assembly. In: Proceeding of the 2nd CIRP Conference on Assembly Technologies & Systems, Toronto, Canada, September 21-23, pp. 407–416 (2008)
9. Simnofske, M., Schöttler, K., Hesselbach, J.: micaboF2 – Robot for Micro Assembly. Production Engineering, XII(2), 215–218 (2005)
10. Rathmann, S., et al.: Sensor-guided micro assembly of active micro systems by using a hot melt based joining technology. Microsystem Technologies 14(12), 1975–1981 (2008)

Grasping and Interaction Force Feedback in Microassembly

Marcello Porta and Marcel Tichem

Department of Precision and Microsystems Engineering (PME),
Delft University of Technology (TU Delft), Mekelweg 2, 2628CD, Delft, The Netherlands
m.porta@tudelft.nl, m.tichem@tudelft.nl

Abstract. The aim of this paper is to define role and added value of force feedback in executing microassembly tasks, as well as to demonstrate microsystem technology based sensing devices which allow registering the relevant forces. In executing microassembly processes useful forces to sense are the grasping force, i.e. the force with which the object is gripped, and the interaction force, i.e. the force resulting from the interaction of the gripped object with the environment. Based on a general assembly process analysis, the requirements for force sensing are defined. Next, devices are shown which allow the identification of the contact forces and object positions.

Keywords: Microassembly, Force sensing, Grasping force, Interaction Force.

1 Introduction

In the assembly of microproducts the feedback of the status of the ongoing process is a key element for the successful completion of the tasks [1], [2], [3]. This feedback can be either visual information or force information. The *grasping force* monitoring (i.e. the force with which the object is gripped) prevents damaging delicate and fragile components (such as micro-lenses; micro-capillary-tubes). The *interaction force* sensing (i.e. the force resulting from the interaction of the gripped object, or the gripper itself, with the environment) is useful for speeding up the process and sometimes necessary for successfully completing the tasks (such as peg-in-hole with low clearance, engagement of micro-gears). Fig. 1 provides some significant examples.

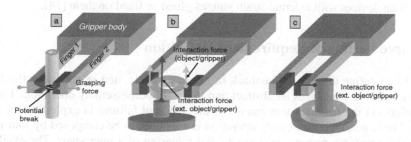

Fig. 1. Operations in which the monitoring of the grasping and interaction forces is important: (a) grasping of a fragile micro-cylinder; (b) peg-in-hole task; (c) interaction between the gripper and a sub assembled product.

S. Ratchev (Ed.): IPAS 2010, IFIP AICT 315, pp. 199–206, 2010.

2 Grasping and Interaction Force Sensing in Microhandling

Many devices have been developed for detecting grasping and interaction forces in microassembly and microhandling.

The grasping force is usually detected by two fingers mechanical micro-grippers equipped with sensors [4], [5], [6], [7], [8], [9], [10]. Some of these grippers are also able to sense interaction force, along one degree of freedom (DOF), perpendicular [4], [5] or parallel [6] to the grasping plane (i.e. the plane defined by the opening and closing of the fingers of the gripper).

The performance of the micro-grippers in terms of stroke, force resolution and maximum force depends on the adopted actuators and sensors, the material of the mechanical structure of the gripper and the mechanical amplification. Usually, there is a trade-off between stroke and force resolution and between maximum force and force resolution. This trade off is evident in the two main types of micro-grippers: silicon-based micro-grippers [5], [6], [9], [10] and hybrid micro-grippers [4], [7], [8]. The silicon micro-grippers usually offer the best resolution while the hybrid ones the best stroke and the maximum grasping force.

Silicon micro-grippers are directly fabricated with integrated actuators (electro-static [9] or thermal [6], [10]) and sensors (piezoresistors [5], [10] or capacitive sensors [6], [9]). In hybrid micro-grippers the mechanical structure of the grippers, the actuators (piezo-elements [4], [8], electromagnets [7] or SMA actuators [8]) and sensors (usually strain gauges [4], [8] glued or fixed to the mechanical structure) are produced separately and then assembled together.

The selection of one of the two main types of microgrippers, their mechanical design, the adopted sensors and actuators depends on the microparts to grasp and the handling applications.

The interaction forces sensing along many DOF (till 6) is normally obtained by devices placed in the interface between the gripper and the manipulator supporting the gripper [11], [12], [13], [14].

The performance of the interaction force sensing devices in terms of force/torque resolution and maximum force/torque depend both on the mechanical structure of the device and the type of sensor used for detect the forces. Usually, the bigger is the maximum detectable force and the worse is the sensing resolution. Silicon-based devices with embedded sensors (piezoresistor [12], capacitive sensors [11], or piezo-electric elements [13]) often present a much better resolution and a lower operational range than devices with external strain gauges glued or fixed on them [14].

3 Force Feedback Requirement Definition

In order to define the role of feedback in microassembly, and in particular the force feedback, Table 1 analyzes an abstract and general microassembly operation. The table points out the feedback that can avoid the different failures in exploiting the sub-tasks. The general microassembly operation is supposed to be composed by four main sub-tasks: grasping, moving, positioning and releasing of a micro-part. The available feedbacks are visual feedback, grasping force feedback and interaction force feedback. For every possible failure, the table reports what is the type of control and the required information to detect the failure. Then, it suggests a type of feedback which is possibly useful for the control system to take an appropriate action for solving the identified failure. For every feedback three different levels of reliability (low, medium, high) are given. This reliability level quantifies the efficacy and the efficiency of any feedback in avoiding the particular failure.

Table 1. Failures and successes related to feedback in the general microassembly operation

SUB TASK	TYPE OF CONTROL	FAILURE	INFORMATION REQUIRED	POSSIBLE FEEDBACK		ACTION REQUIRED
				Type	Reliability	
GRIP part = establish relationship between part and gripper	check if the part is grasped	part not grasped	area of contact part-gripper	visual	low	increase the closing or the force
				grasping force	high	
	check if the part is on the allowable grasping surface	not be able to mount the part because it is grasped in the wrong position	range of area where the part can be grasped	visual (sign on the parts; superimpose on the part the grasping area)	low	move in the right grasping place and re-grasp it
				grasping force (grasping position/force distribution)	high	
	check if the part is grasped with the right orientation	not be able to mount the part	position of part on grasping surface	visual	low	re-grasp the part with the right orientation
				grasping force (grasping position/force distribution)	high	
	check the gripping force	damage of the part (deformation or breaking)	force exerted	grasping force	high	substitute the part
	check unwanted movement of the part (due to opening of the gripper or external force on the part)	1.loosing of the part	presence of the part	visual	high	re-grasp of the part/grasp another part
				grasping force (grasping position/force distribution)	low (high)	
		2. sliding of the part (can cause the impossibility to mount the part)	relative movement between part and gripper	visual	low	
				grasping force (grasping position/force distribution)	low (high)	
MOVE part = part from the feed location to close to the assembly location	check if there is contact of the part/gripper with other elements present in the environment	1.loosing of the part	contact with other element in the environment	visual	medium	move far from the other elements
		2.sliding of the part				
		3.damage of the part or the gripper		interaction force	high	
MOUNT part = bring the part into the final position	check if the part is approaching correctly to the assembly position	damage of part (deformation or breaking)	contact with other element in the environment	interaction force	high	move far from the other elements
	check if the part is correctly positioned in the assembly position	the part is positioned in the wrong position: wrong assembly	position of the part	visual	medium	
RELEASE part = end relationship between gripper and part	check if the gripper is not in contact with the object	the part is not released/it is released in wrong position: wrong assembly	presence of the part	visual	medium	adopt some releasing strategies [15]
				grasping force	medium	
	check if there are unwanted displacements or forces on the object	the part is released in the wrong position	position of the part	visual	medium	adopt some releasing strategies [15]

From Table 1 is evident how the complete presence of feedback (vision, grasping force and interaction force) can be useful or even necessary to successfully carry out the assembly tasks.

Despite visual feedback is potentially able to detect most of the common failures, it does not always ensure a high reliability in terms of efficacy (e.g. to check little movement of the object on the grasping surface) or efficiency (e.g. giving an immediate feedback when the part is not grasped because the gripper has to move far from the grasping position before the part is recognized as not grasped). Furthermore, visual control may no longer be obtainable (e.g. when an obstacle blocks the view or because of the limitation related to the trade off between depth of focus and magnification). Often, the visual control is usually available during the first stages of an assembly operation (i.e. grasping of the parts and moving towards the final location) but not anymore in the final ones (i.e. the mounting and the releasing of the parts) for the narrow and inaccessible spaces in which some microassembly tasks take place.

On the other hand, Table 1 points out that the grasping and the interaction force feedback remain active during the entire assembly process in any assembly locations. Furthermore, these force feedbacks are usually more reliable than the visual feedback because these are more effective and efficient in detecting the failures.

4   Grasping and Interaction Force Challenges

Different challenges in the grasping and interaction force sensing need research efforts to assure the successful execution of many microassembly tasks.

A first requirement is the capability of the gripper to detect *the position and the orientation of the object on the grasping surface*. This information highlights if the object is grasped in the correct place for carrying out the assembly task and if it is safely grasped (Fig. 2). Such capabilities would give the possibility to detect unwanted movements of the object on the grasping surface due to external force.

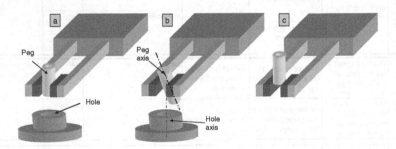

Fig. 2. Tasks in which the knowledge of the grasping position and orientation is important: a) the peg can not be inserted in the hole due its position on the grasping surface; b) the peg orientation does not allow the insertion in the hole because the axis of the peg is not coaxial with the axis of the hole (the gripper must orientate the peg); c) the object is close to the grasping surface edge and risk to be loosen.

A second requirement is the detection of the *grasping force distribution on the object surface*. If the object is not correctly grasped or it has different stiffness and fragility on different locations (Fig. 3), the simple control of the global grasping force can be not enough to avoid breaking or damaging the micro-parts.

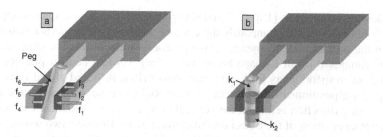

Fig. 3. Examples of situation in which the knowledge of the distribution of the grasping force is important to avoid damaging the micro-parts: a) the grasping force is not exerted uniformly on the micro-parts; b) the micro-parts have difference in stiffness in various location

A third requirement is to *provide the same handling tool with the capability of monitoring both the grasping and interaction force.* This capability would drastically improve tasks such as peg-in-hole insertion of fragile parts with low clearance and engagement of micro-gears. Actually, till now these two forces are usually detected by separate devices as pointed out in Section 2. This means that the gripper with grasping sensing and the device for the interaction force sensing are produced separately. Then the gripper has to be mounted (e.g. by mechanical connection or gluing) on the interaction force sensing device and the interaction sensing device fixed on the manipulator. Since these mounting operations are not simple due to the fragility and the (often) submillimeter dimension of the devices, inaccuracies and offsets affecting the grasping and interaction force sensing performance can arise.

5   New Devices for Position Detection on the Grasping Surface

In order to provide the gripper with the capability of detecting the object grasping position some new sensors have been developed by the authors [16], [17]. The sensors are fabricated with well known silicon based technologies and can be integrated in micro-grippers with IC-compatible process without any post processing assembly tasks.

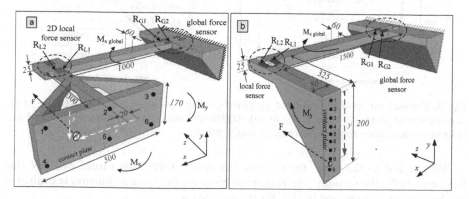

Fig. 4. Schematic drawing of the developed devices for detecting the contact position and the force acting on the contact plate

The devices reported in [16], [17] and shown in Fig. 4 can give microgrippers the capability of detecting simultaneously the contact position and the perpendicular force acting on the contact plate by means of two piezoresistor sensors. The structure of the device is composed of an L-shape beam and a deep vertical contact plate (grasping surface). Piezoresistors pairs for local force sensor (R_{L1}, R_{L2}) and global force sensor (R_{G1}, R_{G2}), are positioned on the beam. The global force sensor detects the contact force F in the z direction acting on the vertical contact plate while the local force sensor the contact position of the object on the contact plate. Both the two sensors consist of four piezoresistors in a Wheatstone bridge configuration. The position sensor of the device shown in Fig. 4a monitors the contact position along both x and y axis while the one of Fig. 4b detects the position along the y axis only.

Both devices were characterized with tests in which either a needle or a microsphere was used to exert a force perpendicular to the contact plate in the points shown in the drafts of Fig. 4 (1-6 in Fig. 4a and 1-9 in Fig. 4b). In the characterization tests the two sensors were supplied with an input voltage of 1V. The stress induced in the two sensors by the force was detected by monitoring the variation of resistance of the piezoresitors; then, the contact position and the global force can be computed.

The results about the relation between the real and the detected contact position are reported in Fig. 5. For the two devices different levels of force (corresponding to different levels of displacement) have been applied in each tested point. The maximum error between real and measured position is 20 μm for the first device (Fig. 5a) and 10 μm for the second one (Fig. 5b). These errors are supposed to be due to: i) the friction force between the object and the contact plate that induces extra stress on the sensor; ii) noise in the signal; iii) resolution of the lock-in amplifiers used to detect the signals.

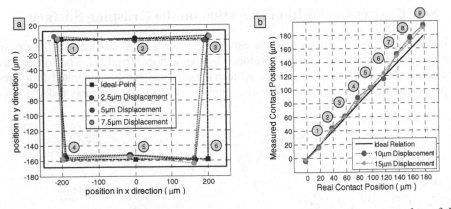

Fig. 5. Comparison between actual and measured contact positions on the contact plate of the devices shown in Fig.4a (a) and Fig.4b (b). Different forces corresponding to different displacements have been applied in each investigated point.

With regard to the force, the devices can detect a maximum force of 4mN (the first) and 3mN (the second). The maximum error for the force estimation is evaluated in 15% for the first one and in 10% for the second one.

6 Conclusions and Future Developments

In this paper the advantages of a complete force feedback (grasping and interaction force) in microassembly tasks have been described and discussed. Some grasping and interaction force challenges have been highlighted and devices developed by the authors that tackle some of these challenges shown as well.

Future development aims at implementing the force feedback improvement described in the paper. The developed sensors for the detection of the position of the grasped micro-object will be integrated and tested in microgrippers. Gripper solutions with both grasping force sensing and interaction force detection along many DOF will be studied and developed as well.

Acknowledgement

The authors wish to thank the PME technical staff and the DIMES IC Process Group for their precious help in the fabrication and tests of the devices. Special thanks are due to Dr. Jia Wei and Prof. Lina Sarro for their support and the useful discussions in the research. Our gratitude to MicroNed program (www.microned.nl) for financing this research.

References

1. Rabenorosoa, K., Haddab, Y., Lutz, P.: A Low Cost Coarse/Fine Piezoelectrically Actuated Microgripper with Force Measurement Adapted to Eupass Control Structure. In: Proc. of IPAS 2008, Chamonix (France), February 10-13, pp. 235–242 (2008)
2. Probst, M., Hürzeler, C., Borer, R., Nelson, B.J.: A Microassembly System for the Flexible Assembly of Hybrid robotic Mems Devices. Int. Journal of Optomechatronics 3(2), 69–90 (2009)
3. Sieber, A., Valdastri, P., Houston, K., Eder, C., Tonet, O., Menciassi, A., Dario, P.: A novel haptic platform for real time bilateral biomanipulation with a MEMS sensor for tri-axial force feedback. Sensors and Actuators A 142, 19–27 (2008)
4. Voyles, R.M., Hulst, S.: Micro/macro force-servoed gripper for precision photonics assembly and analysis. Robotica 23(4), 401–408 (2005)
5. Keller, C.G., Howe, R.T.: Hexsil Tweezers For Teleoperated Microassembly. In: Proc. of 10th IEEE Int. Workshop on Micro Electro Mechanical Systems (MEMS 1997), Nagoya (Japan), January 26-30, pp. 72–77 (1997)
6. Kim, K., Liu, X., Zhang, Y., Cheng, J., Yu Wu, X., Sun, Y.: Elastic and viscoelastic characterization of microcapsules for drug delivery using a force-feedback MEMS microgripper. Biomedical Microdevices 11(2), 421–427 (2009)
7. Kim, D.H., Lee, M.G., Kim, B., Sun, Y.: A superelastic alloy microgripper with embedded electromagnetic actuators and piezoelectric force sensors: a numerical and experimental study. Smart Material and Structures 14, 1265–1272 (2005)
8. Kyung, J.H., Ko, B.G., Ha, Y.H., Chung, G.J.: Design of a microgripper for micromanipulation of microcomponents using SMA wires and flexible hinges. Sensors and Actuators A 141, 144–150 (2008)

9. Beyeler, F., Neild, A., Oberti, S., Bell, D.J., Sun, Y., Dual, J., Nelson, B.J.: Monolithically Fabricated Microgripper with Integrated Force Sensor for Manipulating Microobjects and Biological Cells Aligned in an Ultrasonic Field. Journal of MEMS 16(1), 7–15 (2007)

10. Chu Duc, T., Lau, G.K., Creemer, J.F., Sarro, P.M.: Electrothermal microgripper with large jaw displacement and integrated force sensors. Journal of MEMS 17, 1546–1555 (2008)

11. Beyeler, F., Muntwyler, S., Nelson, B.J.: Design and Calibration of a Microfabricated 6-Axis Force-Torque Sensor for Microrobotic Applications. In: 2009 IEEE Int. Conf. on Robotics and Automation (ICRA 2009), Kobe (Japan), May 12-17 (2009)

12. Viet Dao, D., Toriyama, T., Wells, J., Sugiyama, S.: Silicon Piezoresistive Six-Degree of Freedom Micro Force-Moment Sensor. Sensors and Materials 15(3), 113–135 (2003)

13. Shen, Y., Xi, N., Lai, K.W.C., Li, W.J.: A novel PVDF microforce/force rate sensor for practical applications in micromanipulation. Sensor Review 24(3), 274–283 (2004)

14. http://www.ati-ia.com/products/ft/ft_models.aspx?id=Nano17 (last access 14 October 2009)

15. Fantoni, G., Porta, M.: A critical review of releasing strategies in microparts handling. In: Proc. of IPAS 2008, Chamonix (France), February 10-13, pp. 223–234 (2008)

16. Porta, M., Wei, J., Tichem, M., Sarro, P.M., Staufer, U.: Vertical Contact Position Detection and Grasping Force Monitoring for Microgripper Applications. In: Proc. of IEEE Sensors 2009, Christchurch (New Zeeland), October 25-29 (2009)

17. Wei, J., Porta, M., Tichem, M., Sarro, P.M.: A Contact Position Detection and Interaction Force Monitoring Sensor for Microassembly Applications. In: Proc. of Transducer 2009, Denver (CO, USA), June 21-25, pp. 2385–2388 (2009)

Low Voltage Thermo-mechanically Driven Monolithic Microgripper with Piezoresistive Feedback

Vladimir Stavrov[1,*], Emil Tomerov[1], Chavdar Hardalov[2], Daniel Danchev[3], Kostadin Kostadinov[3], Galina Stavreva[1], Evstati Apostolov[2,4], Assen Shulev[3], Anna Andonova[2], and Mohammed Al-Wahab[5]

[1] AMG Technology Ltd., Industrial Zone Microelectronica, Botevgrad, Bulgaria
[2] Sofia Technical University, Sofia, Bulgaria
[3] Institute of Mechanics and Biomechanics, BAS, Sofia, Bulgaria
[4] Micro Plus - Apostolov Ltd., Sofia Bulgaria
[5] University Otto-von-Guiricke, Magdeburg, Germany
vs@amg-t.com

Abstract. Pick-and-place of micro/nano sized objects means handling of very tiny and very different in properties objects having specific behavior. Besides formal requirements of assembly processes, the tools for controllable manipulation with these objects should not affect the examined micro/nano environment, i.e. should be "small and passive" in any sense. Despite of the recent progress, most available micro-grippers are still suffering of high voltage power supply required, short lifetime, low detection sensitivity and high price.

Prototypes of a newly designed micro-gripper, having advantages over the existing analogues, have been developed, experimentally studied, and presented in this paper. The envisaged microgripper is of normally-closed type with thermo-mechanically driven actuator and piezoresistive arm-displacement feedback. The thermo actuator is placed between gripper's arms and consists of double-folded highly-doped compliant silicon beam. As low average voltage vs. arm displacement value as 1V/μm, was experimentally measured.

Keywords: MEMS, micro-gripper, piezoresistive detection.

1 Introduction

Formal requirements and design trade-off for pick-and-place micro assembly for microgrippers are well described elsewhere [1]. One of the important characteristic features is to provide feedback during manipulation with micro/nano sized objects and to achieve handling of very small and different in properties objects having specific behavior, respectively. Besides controllable manipulation with different in properties objects, these tools should not affect the examined micro/nano environment, i.e. should be "small and passive" in any sense. That is why a variety of grippers dedicated for different applications have been developed during the recent decade [2], [3]. Despite of the progress, most available microgrippers are still suffering drawbacks

---

* Corresponding author.

S. Ratchev (Ed.): IPAS 2010, IFIP AICT 315, pp. 207–214, 2010.
© IFIP International Federation for Information Processing 2010

like: high voltage power supply required, short lifetime, low detection sensitivity and high price. In some cases embedded feed-back sensors are integrated but in general they are non-linear and have low detection sensitivity.

Microgrippers with thermo-mechanical actuation have been considered as one of the promising solutions for bio-medical applications [4]. At the same time, due to the high dissipated power in small-size tools, this type of microgrippers suffers of additional problems like: 1) it is easy to overheat the handled object if non-proper design is used; 2) due to elevated temperatures of operation there is a feed-back sensor parameter drift.

2 Design Considerations of Thermo-mechanically Driven Monolithic Microgripper

To overcome the above mentioned obstacles in further development of monomorph compliant microgripper, subject of this paper, following considerations are taken into account:

- in order to manufacture prototypes of a gripper with embedded feed-back sensing, a single-crystal silicon raw material was preferred. A piezoresistive sensing method of arms deflection was selected. Piezoresistors are self-aligned to sidewalls of both gripper's arms and are located close to their fixed ends, where the stress due to deformation reaches it's maximum value. Thus, a maximal sensitivity of the deflection sensing was achieved;

- it was considered as a very important factor to get low voltage durable devices, so electrostatic and piezoelectric actuation options were rejected. Thermo-mechanical actuation was chosen because it has several advantages like: low-voltage, low-power supply, it is step-less and hysteresis-less method, no technology complication due to the actuator integration, etc.

- power on-time and total heat dissipation, respectively, have been reduced via choosing normally closed type of a microgripper. Further, in order to provide simple layout, a design with thermo-actuating elements placed between gripper's arms has been considered for experiments reported;

- since gripper's arms are moving in the plane of substrate, heating element should be designed to avoid off-plane bending. That's why a monomorph actuator element was preferred. In this particular case a highly doped diffusion heater was used for thermo-actuator element. Since the coefficient of thermal expansion of silicon is as low as $4.2 \cdot 10^{-6}$ °C^{-1}, the length of the heated element should be maximized in order to achieve sufficient end-effector opening range. In the current embodiment the length of the double-folded compliant mechanism is 425 µm. Furthermore, a specific geometry, providing mechanical amplification of thermally-driven displacement, has been used;

- any overheating of the handled object is further suppressed by placing the monomorph actuator at a maximal distance of gripper jaws and, additionally, griper's arms were designed to have as large as possible surface area for intense heat dissipation.

3 Description and Principle of Operation

Microgripper shown in Fig. 1 consists of a body *1* and two symmetrical arms, *2* and *2'* respectively, each of them having thickness between 12 μm and 15 μm and located on the top surface of the body at a distance of 520 μm from each other. Arms have also deformable parts *3* and *3'* where piezoresistors *4* and *4'* are embedded on their sidewalls thus sensing lateral deflections of the arms. The distance between two jaws of the gripper *5* was set to about 10 μm. The arms have been provided with square-shaped hollows *6* which give extra heat dissipating surface and reduce heat transfer from the body to the jaw's area, where the manipulated object have to be in contact with the gripper. Special care was taken to get stiff arm-construction and to prevent it's bending elsewhere, besides deformable parts *3* and *3'*, so arm-elements *7* have been added.

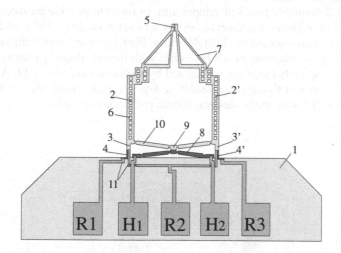

Fig. 1. Schematic layout of thermo-mechanically driven monolithic micro-gripper

When DC current in the range of 5-30 mA is supplied to terminals H_1 and H_2 of heavily diffusion-doped silicon heating element *8* with resistivity of 200-250 Ω, the Joule heat causes temperature rise. The heated compliant element having 425 μm length and sheet resistance of 6 Ω/□, consists of three parts with flexible joins (hinges): two symmetric beams *8* and a central part *9*, with both outer ends of this compliant mechanism being fixed. Due to compliant design, thermal expansion is converted into part *9* translation movement at off-substrate edge direction and its movement is transferred by transmission beams *10* to gripper arms causing jaw's opening. Regulating the heating of the compliant element, one can set the gripper opening at desirable width. Once the power supply is interrupted, due to heat exchange with environmental media, griper comes back to closed position.

Embedded piezoresistors, which are sensitive to in-plane movement of gripper arms, are located into deformable area of the arms' fixed end. These resistors are connected with metal tracks *11* to R_1, R_2 and R_3 pads. Thus, the end-effector movement causes a changing piezoresistor value. In this way, arm-end position sensing and controllable manipulation of the objects could be achieved. The maximum value of the gripping force is defined by the stiffness of the deformable parts *3* and *3'* of the arms, but a range of intermediate forces is achieved via permanent heating during object handling, as well.

4 Microgripper Prototype Processing

The monolithic gripper prototypes have been micro machined by using double-side polished n-type (100) silicon wafers with a resistivity of 4-6 Ω.cm and TTV < 2µm. Fig. 2 shows a drawing of gripper arm cross-section. Gripper's micromechanical elements are made by means of combination of dry surface and wet bulk micro-machining processes as an integral part of recently developed technology for lateral actuated MEMS manufacturing. P-type diffusion resistors *4* and *4'* are self-aligned to sidewalls of deformable part *3* of gripper arm. In order to provide piezoresistive properties of diffusion layers, the sheet resistivity was set at range of 250 ± 20 $\Omega/\square$ and the resistors have been oriented in [110] direction. Piezoresistor's non-rectifying contacts are provided by overlapping of it's both ends with heavily doped p+-areas *13* and this structure is completely electrically insulated by silicon dioxide layer *14*. A contact via *15* have been opened through the insulating layer to make available galvanic contact of metal tracks *11* with above mentioned both piezoresistor's ends.

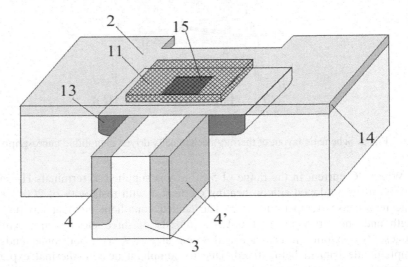

Fig. 2. Cross-section of gripper arm with sidewall embedded piezoresistors on deformable part

5 Results

5.1 Prototypes of Thermo-mechanically Driven Monolithic Microgripper

Fig. 3 shows a Scanning Electron Microscopy photograph of microgripper prototype which is dedicated to bio-medical applications. The gripper has been implemented as a monomorph compliant mechanism and special care was taken for reducing temperature in the area of jaws *5*. Despite high temperatures required for thermomechanical actuation, this particular design provides high heat resistance along the grippers' arms and large sidewall area for heat exchange with environmental media. At the same time the design provides sufficient mechanical stiffness of the microgripper arms *2(2')*.

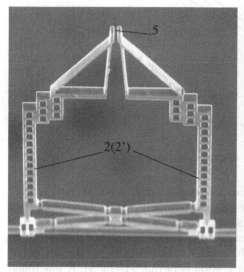 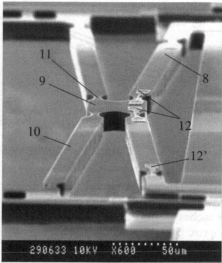

Fig. 3. SEM photo of thermo-mechanically driven monolithic gripper

Fig. 4. SEM photo of micro-gripper's compliant mechanism with thermo-mechanical actuator

The scanning electron micrograph (Fig. 4) shows the above mentioned compliant mechanism in more details: right part is the heavily doped heater *8* and left part is the translation transmitting beams *10*. Since cross-section area of heating element *8* is changing along it's length, the resistance of the flexible joints (hinges) *11* is higher than resistance in the rest parts of it. In order to avoid overheating in these areas, additional metal shunts: central *12* and outer *12'* are provided. Shunts are made of the same thin film material as metal wiring and for these particular prototypes metal layer was 0.8 μm thick aluminum (Al).

5.2 Electrical Measurements

To determine the deflection of the gripper's arms, the heater is power supplied by a DC current as shown in Fig. 1. Heating power was determined by measuring the voltage drop over the monolithic heater.

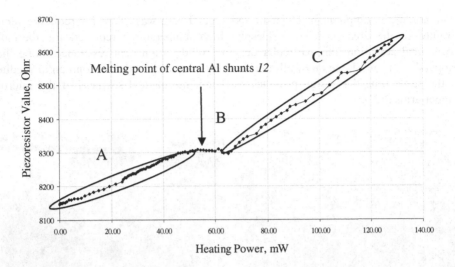

Fig. 5. Piezoresistor value change vs. Heating power

Fig. 5 shows a typical plot of measured piezoresistor values vs. applied heating power. Three distinguished regimes are observed depicted in the three distinctive branches of the curve. The first one (A) is when the heating power is between 0 to approx. 50 mW which shows a linear increasing resistance. Similar behavior could be observed for the third range (C) between 70 and 140 mW. For the middle part (B) of heating power between 50 and 70 mW a saturation behavior was found. It was found that latest range corresponds to heating of compliant central part 9 (Fig. 4) up to 660 °C, and melting of the central aluminum shunts *12*. Optical microscopy observations confirm this hypothesis but more detailed studies of this region of heating power will be conducted further.

5.3 Mechanical Behavior of Micro-gripper

The opened and closed jaws of the gripper are visualized on Figures 6 and 7.

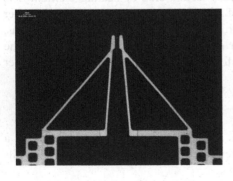
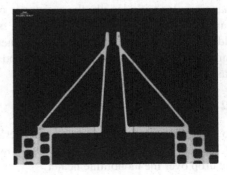

Fig. 6. Photo of closed microgripper jaws   **Fig. 7.** Photo of opened microgripper jaws

6 Thermal Behavior: FE Simulation and Experimental Results

Experimental measurement of temperature distribution over the monolithic micro-gripper surface was not available within present study, thus the computer simulation 3D model has been derived and extruded from the photolithography masks. The control equations consist of the steady-state Conductive Media DC (emdc), General Heat Transfer (htgh) and Solid Stress-Strain (smsld) from the Model Navigator tool of COMSOL Multiphysics. The relevant boundary conditions have been selected and adapted to the experimental results shown in Fig. 8. One observes that saturation temperature within the region of gripper's jaws, shown in Fig. 9 could reach 460°K for as low heating power as 20mW. Thus, bio-compatible manipulations (T< 315°K) with present microgripper are limited to short power on-time cases, only.

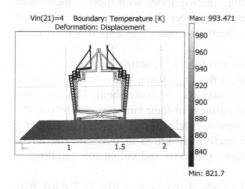

Fig. 8. Temperature distribution over the micro-gripper at typical conditions

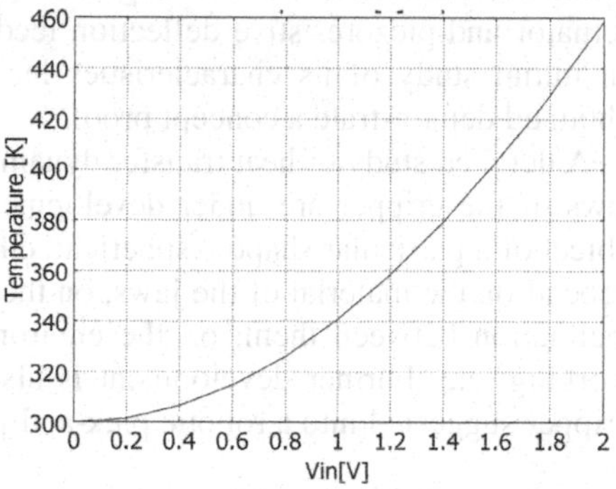

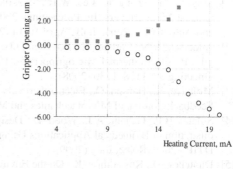

Fig. 9. Micro-gripper jaws' saturation tempe-rature as a function of applied heating power

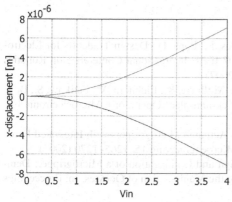

Fig. 10. Micro-gripper jaws opening as a function of applied current (simulation)

Fig. 11. Micro-gripper arms opening vs. heating current (experimental)

Figs. 10 and 11 represent a comparison of the computer simulation data for microgripper's arms displacement -- see Fig. 10, with the experimentally measured results shown in Fig. 11. One concludes a good agreement between the values of microgripper jaws opening measured with the values predicted via computer simulation.

Finally, let us stress that the experimentally measured average value of the coefficient between the electrical voltage applied and each arm displacement is 1 μm/V, which demonstrates that indeed low voltage operation control of the action of the monolithic thermo-mechanically driven micro gripper has been achieved within the realization suggested.

7 Conclusion

In the current article a low voltage monolithic microgripper with thermo-mechanical actuator and piezoresistive deflection feedback has been developed and the results of an initial study of its characteristics have been reported. The experimental results obtained demonstrate a concept proof.

A detailed study of heat transfer dynamics and forces [5] acting between and on the jaws of the gripper are under development in the case when there is or there is no object of a particular shape – spherical, ellipsoidal, etc., between them. This force will depend on the material of the jaws, on the roughness of their surfaces, on the possible inclination between them, on the environmental conditions in which the gripper is working, etc. Further development is also performed aimed to integrate the microgripper suggested into a robotic pick-and-place system for individual cell handling.

Acknowledgements. Authors are thankful to Mr. J. Kirov and Mr. P. Pavlov from AMG Technology Ltd. for processing the prototype and measurement of samples.

The partial financial support of grant TK171/08 of Bulgarian NSF is gratefully acknowledged.

References

[1] Mayyas, M., Zhang, P., Lee, W.H., Shiakolas, P., Popa, D.: Design Tradeoffs for Electrothermal Microgrippers. In: Proceedings 2007 IEEE International Conference on Robotics and Automation, Roma, Italy, April 10-14 (2007)

[2] Andersen, K.N., Carlson, K., Petersen, D.H., Mølhave, K., Eichhorn, V., Fatikow, S., Bøggild, P.: Electrothermal Microgrippers for Pick-and-place Operations. Microelectronic Engineering 85, 1128–1130 (2008)

[3] Mølhave, K., Hansen, O.: Electro-thermally Actuated Microgrippers with Integrated Force Feedback. Journal of Micromechanics and Microengineering 15, 1265–1270 (2005)

[4] Solano, B.P., Gallant, A.J., Wood, D.: Design and Optimization of a Microgripper: Demonstration of Biomedical Applications Using the Manipulation of Oocytes. In: Proceedings DTIP - 2009, Rome, Italy (2009)

[5] Dantchev, D., Kostadinov, K.: On the Environmental Influence on the Force Between Two Metallic Plates of a Gripper Immersed in a Nonpolar Fluid: the Role of the Temperature and the Chemical Potential. In: Proceedings of the 4M/ICOMM International Conference on Multi-Material Micro Manufacture (4M/ICOMM 2009), Karlsruhe, Germany, September 23-25, pp. 297–300 (2009)

Improvement of Robotic Micromanipulations Using Chemical Functionalisations

Jérôme Dejeu, Patrick Rougeot, Michaël Gauthier, and Wilfrid Boireau

FEMTO-ST Institute, UMR CNRS 6174 - UFC/ENSMM/UTBM,
24 rue Alain Savary, 25000 Besançon, France
Firstname.surname@femto-st.fr

Abstract. Robotic microhandling is disturbed by the adhesion phenomenon between the micro-object and the grippers. This phenomenon is directly linked to both the object and the gripper surface chemical composition. We propose to control adhesion by using chemical self-assembly monolayer (SAM) on both surfaces. Previous distance-force measurements done with AFM have shown that the liquid pH can be used to modify the adhesion and created repulsive force between the gripper fingers and the micro-objet. This paper shows the correlation between the force distance distance measurements and the micromanipulation tasks using chemically functionalized grippers.

Keywords: Microhandling, Surface functionalisation, Aminosilane, Adhesion, Pull-off forces.

1 Introduction

The assembly of microsystems is a great challenge because of the microscopic size of the components. In fact, the major difficulty of micro-assembly comes from the micro-objects' behaviour which depends on surface forces [1-2]. The manipulation of a micro-object requires handling, positioning, and releasing without disturbances of the surface forces such as electrostatic forces, van der Waals forces or capillary forces. The release is the most critical phase which is usually hindered by adhesion.

Several methods have been proposed in the last ten years to improve micromanipulation. The first approach consists in using non-contact manipulation like laser trapping or dielectrophoresis. These manipulation methods are not disturbed by adhesion but the blocking force stays low and thus cannot be easily applied microassembly. The second approach deals with contact manipulation, where the adhesion is directly used for manipulation or strategies, is used to reduce adhesion between end-effectors and objects.

In this article, we propose a new contact handling system based on a chemical control of the surface forces between the object and the gripper. The objective is to control the adhesion force or to create a repulsive force to guarantee a reliable release. Now, the surface properties of a material can be controlled by surface functionalisation in a liquid by modifying the pH. The charge density on functionalized surfaces is effectively linked to the pH. The microhandling principle based on chemical switching surface is presented in figure 1. The grasping can be done at pH 1 where the surface charge on the

S. Ratchev (Ed.): IPAS 2010, IFIP AICT 315, pp. 215–221, 2010.

gripper and the object induces an attractive force. In order to release the object, the pH is modified to a second value pH 2 where the object charge is changing. The electrostatic force becomes repulsive and the object is released.

This article deals with comparative handling operations performed with different chemical functionalisations at different pH.

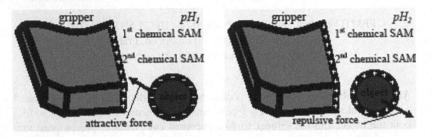

Fig. 1. Principle of the Robotic Microhandling controlled by Chemical Self Assembly Monolayer (SAM)

2 Chemical Functionalisation

2.1 General Principles

The microhandling method proposed is based on two chemical functions: amine and silica. In one hand, the amino group is in state NH_2 in basic pH and in NH_3^+ in acidic pH. In the other hand, the silica surface charge in water is naturally negative, excepted for very acidic pH, where the surface is weakly positive [3].

The surface functionalisation of both object and gripper can be obtained by different methods. The two most important methods are the molecules grafting on the surface (covalent bond between the substrate and the molecules) or the polyelectrolyte physisorption (polyelectrolyte with positive or negative charges) [4].

We chose to investigate the first method because, firstly, it generates covalent bond between substrate and molecules. These molecules must contain silanol, thiol, azide, allyl or vinyl groups in one extremity. These molecules have to be used in organic solvent such as toluene, acetone, methanol, ethanol, etc. The silanol creates a Si-O-Si bond with the silica substrate while allyl or vinyl generates Si-O-C (or Si-C) bond and the acide groups produce Si-N bond [5]. The second reason is that the layer created by silanisation did not exhibit any signature of degradation when stored in an airtight container for 18 months [6], and was stable up to a temperature near 350°C [7] and when washed using 1% detergent solution, hot tap water or organic solvents and aqueous acid at room temperature [7].

2.2 Chemical Components

The chemical compounds (APTES) used to surface functionalisation are amine functions NH_2 which can protonated or ionised to NH_3^+ according to pH. In acidic pH, the amine is totally ionised, then the ionisation decreases and is null in basic pH (between pH 9 and 12).

The silanes (APTES), ethanol, sodium chloride (NaCl), sodium hydroxyde (NaOH) and chlorydric acid (HCl), were purchased from Sigma Aldrich.

2.3 Functionalisation Mechanisms: Grafted Silanes

The self assembled monolayers formation mechanism during silanization process takes place in four steps [6]. The first step is physisorption, in which the silane molecules get physisorbed at the hydrated silicon surface. In the second step, the silane head-groups comes close to the substrate hydrolyse, in the presence of the adsorbed water layer on the surface, into highly polar hydroxysilane -Si(OH). These polar Si(OH) groups form covalent bonds with the hydroxyl groups on SiO_2 surface (third step). At least, condensation reaction (release of water molecules) appears between silanol functions and neighbor molecules. Self assembly is driven by lipophilic interactions between the linear alkane moieties.

During initial period, only a few molecules will adsorb (by steps 1-3) on the surface and the monolayer will definitely be in a disordered (or liquid) state. However, at long term, the surface coverage eventually reaches the point where a well-ordered and compact (or crystalline) monolayer is obtained (step 4).

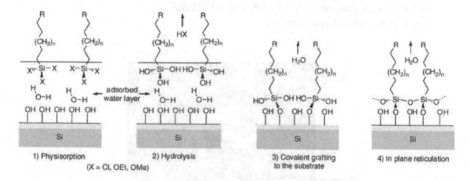

Fig. 2. Steps involved in the mechanism of SAM formation on a hydrated silicon surface [6]

2.4 Characterisation of the Fonctionalised Surfaces

Force measurements were performed in order to characterize the functionalisations [8-9]. Force-distance curves were performed using a stand-alone SMENA scanning probe microscope (NT-MDT).

The force measurement performed on this Atomic Force Microscope (AFM) is based on the measurement of the deformation of the AFM cantilever with a laser deflection sensor. The silicon rectangular AFM cantilever, whose stiffness is 0.3N/m, was fixed and the substrate moved vertically. As the applicative objective of this work is to improve reliability of micro-object manipulation, interactions have been studied between a micrometric sphere and a plane. Measurements were in fact performed with a cantilever where a borosilicate sphere (r_2=5 μm radius) was glued (company Novascan Technologies, Ames, USA).

All measurements were done at the driving speed of 200 nm/s to avoid the influence of the hydrodynamic drag forces [10]. For each sample, nine measures were done in different points. The repeatability of all the pull-off and pull-in forces values was better than 10 %.

3   Micromanipulation System

The impact of the functionalisation has been tested on a robotic micro-assembly device composed of a robotic structure, optical microscope, a piezogripper.

Our current robotic micro-assembly device is presented in figure \ref{Fig:Station}. Actuation is divided into two groups which have 3 degrees of freedom (DOF). The first one allows displacement of the substrate, where microparts are placed. Two linear and one rotation DOFs are available in the horizontal plane. The second group is a 'robotic arm', composed of three linear DOFs and two rotation DOFs which enable displacements and rotations of the microgripper.

Moreover, the microgripper is based on a 4 DOFs piezoelectric actuator with silicon finger tips [11]. This gripper induces motions with a great resolution (few tenth of nanometers).

Experiments have been done in wet medium with a functionalised object and :

- end-effectors grafted with APTES and APDMES or
- non-functionalised end-effectors.

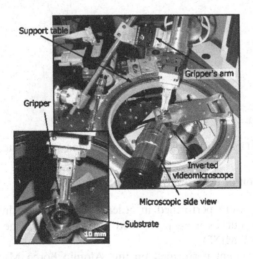

Fig. 3. Microassembly robotic device

Microscopical vision is provided by two videomicroscopes. As the volume above the micromanipulation plane is dedicated to microgripper movement, an inverted microscope LEICA DM-IRBE is used. It enables micro-assembly in liquid medium, whose general interest is synthesized in [2]. A second view for teleoperated operations is given by a side videomicroscope.

4 Micromanipulation Results

4.1 Non- functionalized Objects

The micromanipulation tasks are often disturbed by the adhesion force between the grippers and the micro-object. Indeed, with the AFM, we measure a pull-off greater than 1 μN (Figure 4). This important adhesion force is confirmed by the micromanipulation task with the micromanipulation system. The glass sphere between 1 and 45 μm stay attach to the end-effector of the gripper when we open the gripper (Figure 5). The glass spheres are as glued on the end-effector whoever the glass size. Indeed, in the range from 1 to 45 μm, we observe the same phenomena, no release of the glass when we open the gripper.

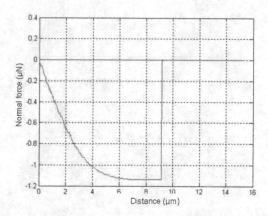

Fig. 4. Force-distance curve between borosilicate cantilever and silica surface (spring constant 0.3 N/m)

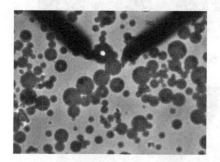 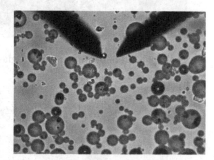

Fig. 5. Glass sphere adhesion during micromanipulation task. Rigth: small sphere near 3 μm$ and Left: big sphere near 45 μm.

The adhesion can be cancelled by functionalisation of the grippers and the spheres.

4.2 Functionalised Objetcs

In a previous paper, we have shown a repulsion between the surface and cantilever when both are functionalised by an aminosilane and this whatever the pH of the solution. In this paper, illustration of this phenomena is shown in micromanipulation tasks. As the repulsive force has a high interaction distance (typicallytens of micrometers) [9], this effect can be used to manipulate the object without contact with the end-effector. Indeed, when the functionalized end-effector is approached near to the functionalized object, the object is moving behind the end-effector (Figure 6). When the ball is caught, the sphere release is easy because of the repulsive force. This release is illustrated Figure 7. In this Figure, we observe that the adhesion does not disturb the release.

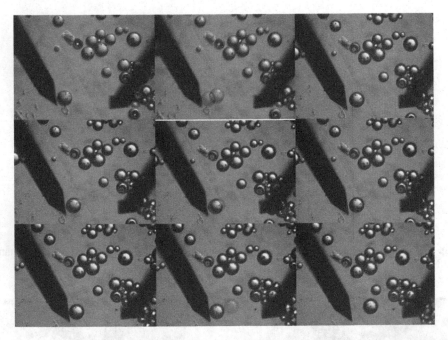

Fig. 6. Repulsion of the functionalised glass sphere when a functionalised SiFit approach to its

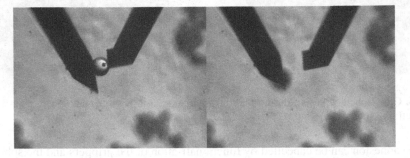

Fig. 7. Release of functionalised glass sphere when the functionalised gripper open

5 Conclusion

In this paper we have observed the adhesion behaviour between two functionalised surfaces and between two surfaces. We have shown the correlation between the force-distance measurement and the micromanipulation tacks. The liquid medium and the chemical functionalisation can be used to control the surface properties and the micro-object release. As adhesion is the current highest disturbance in micromanipulation, functionalisation is a promising way to improve micro-object manipulation in the future.

Acknowledgement

This work was supported by the EU under HYDROMEL contract NMP2-CT-2006-026622 : Hybrid ultra precision manufacturing process based on positional- and self-assembly for complex micro-products, and by the French National Agency (ANR) under NANOROL contract ANR-07-ROBO-0003: Nanoanalyse for micromanipulate.

References

[1] Lambert, P.: Capillary Forces in Micro-assembly. Springer, Heidelberg (2008)
[2] Gauthier, M., Régnier, S., Rougeot, P., Chaillet, N.: Forces analysis for micromanipulations in dry and liquid media. Journal of Micromechatronics 3, 389–413 (2006)
[3] Dove, P., Craven, C.: Surface charge density on silica in alkali and alkaline earth chloride electrolyte solutions. Geochimica et Cosmochimica Acta 69, 4963–4970 (2005)
[4] Dejeu, J., Lakard, B., Fievet, P., Lakard, S.: Characterization of charge properties of an ultrafiltration membrane modified by surface grafting of poly(allylamine) hydrochloride. Journal of Colloid and Interface Science 333, 335–340 (2009)
[5] Wang, J., Guo, D., Xia, B., Chao, J., Xiao, S.: Preparation of organic monolayers with azide on porous silicon via sin bonds. Colloids and Surface A: Physicochemical and Engineering Aspects 305, 66–75 (2007)
[6] Wasserman, S., Tao, Y., Whitesides, G.: Structure and reactivity of alkylsiloxane monolayers formed by reaction of alkyltrichlorosilanes on silicon substrates. Langmuir 5, 1074–1087 (1989)
[7] Ulman, A.: An Introduction to Ultrathin Organic Films From Langmuir-Blogett to Self-Assembly. Academic Press, London (1991)
[8] Dejeu, J., Rougeot, P., Gauthier, M., Boireau, W.: Adhesions forces controlled by chemical self-assembly and ph, application to robotic microhandling. ACS Applied Materials & Interfaces 1, 1966–1973 (2009)
[9] Dejeu, J., Rougeot, P., Gauthier, M., Boireau, W.: Robotic submerged microhangling controlled by ph switching. In: Proc. of the IEEE IROS 2009 conference, St Louis (2009)
[10] Vinogradova, O., Yakubov, G.: Dynamic effects on force measurements. 2. lubrication and the atomic force microscope. Langmuir 19, 1227–1234 (2003)
[11] Agnus, J., Hériban, D., Gauthier, M., Pétrini, V.: Silicon end-effectors for microgripping tasks. Precision Engineering 33, 542–548 (2009)

Positioning, Structuring and Controlling with Nanoprecision

Regine Hedderich[1,2], Tobias Heiler[2,3], Roland Gröger[2,3],
Thomas Schimmel[2,3], and Stefan Walheim[2,3]

[1] Network NanoMat
[2] Institute of Nanotechnology, Forschungszentrum Karlsruhe
Karlsruhe Institute of Technology (KIT), Northern Campus
76344 Eggenstein Leopoldshafen, Germany
[3] Institute of Applied Physics
and Center for Functional Nanostructures (CFN)
Universität Karlsruhe / Karlsruhe Institute of Technology (KIT), Southern Campus
76131 Karlsruhe, Germany
regine.hedderich@kit.edu
http://www.nanomat.de

Abstract. Key industries such as the automotive, electronic, medical and laboratory technical industries have continually rising demands for precise manufacturing, handling and control techniques. This is true for the manufacture of injection nozzles for engines as indeed also for the irradiation of extremely fine wafer structures and in the field of scanning probe microscopy. Some examples from research and their industrial application which have been made available by NanoMat Network Partners will be highlighted in this presentation. For example the spontaneous structure formation at the nanoscale and the application as anti-reflection layers.

Keywords: Nanotechnology, AFM, lithography, polymer blends.

1 Applications

The spontaneous structure formation in thin polymer films during spin coating can be used to define 2D and 3D chemically patterned surfaces, which feature complex and hierarchically organized structure motifs - both on the nanometer and on the micrometer scale. Investigations made by in situ light scattering during the typically five seconds of film formation give insight into the structure formation process [1]. An example of derivatives of these structures are bio-functionalizable structures with laterally defined spots of only 10 nm in diameter which are fabricated by the combination of polymer blend lithography and block copolymer nanolithography. By the use of micro-contact printing and scanning probe lithography - the self-organisation of the resulting polymer pattern can be controlled such that well ordered layout-defined structures can be achieved [2]. Besides e-beam lithgography and printing

S. Ratchev (Ed.): IPAS 2010, IFIP AICT 315, pp. 222–226, 2010.
© IFIP International Federation for Information Processing 2010

techniques like micro-contact printing laser interference lithography can be used to generate appropriate Surface energy patterns on the substrates to guide the structure formation during spin-coating [3],[4]. But also a purely isotropic phase separation which take place on unstructured surfaces can lead to new physical properties. For instance, if we remove one of the two polymers of a morphology with a lateral length scale below 200 nm, we are left with a nanoporous film. By tuning the composition of the polymer solution, the refractive index of our films can be adjusted in a range from 1.6 down to 1.05. This films can be used as high-performance anti-reflection coatings with outstanding optical properties [5].

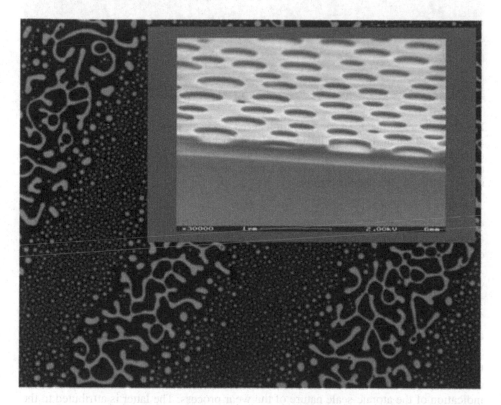

Fig. 1. Optical Microscopy image and SEM images of hierarchically structured surfaces made by polymer blend lithography. A complex structure formation process produces three characteristic correlation lengths at the same time. The diameter of holes in the upper picture is about 200 nm (scale bare: 1 μm). These polymer structures can be produced homogeneously on large samples and transferred into metal- and semiconductor structures.

Further researches in our Nanotechnology is controlled structuring of mica surfaces with the tip of an atomic force microscope by mechanically induced local etching. Here we show a reproducible structuring of surfaces of muscovite mica with the tip of

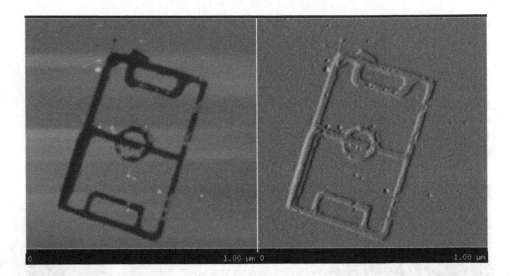

Fig. 2. Nanostructuring using our AFM-tip-induced mechanical etching process. A structure of 300 nm x 700 nm in size was written on a mica surface. The groove width (FWHM) is 30 nm and the depth is about 3 nm. The AFM images (image size 1 μm x 1 μm) of the resulting structures were taken in the constant amplitude tapping mode of the microscope: (left) topographic image: (right) amplitude deviation image.

an atomic force microscope under ambient conditions. By repeated scanning of the tip along a predefined pattern on a cleaved mica surface, mechanically induced etching can be observed on the atomic scale.

Line widths down to 3 nm were achieved, while at the same time patterns on a wide range of length scales between 5 nm and 100 μm are generated reproducibly [6]. This can be explained by abrasive wear on the atomic scale due to sliding friction. The experiments allow the study of tribochemistry and abrasive wear on the atomic and molecular scale. At the same time, they represent an approach for high-precision structuring of surfaces within a wide range of length scales. The structuring of the topographic patterns was performed under ambient conditions. Neither debris particles nor ecrystallization of debris products on the surface was found. The former is an indication of the atomic-scale nature of the wear process. The latter is attributed to the fact that the experiments are performed under ambient conditions, including the presence of an adsorbed nanoscalewater layer. The experiments allow the study of abrasive wear under ambient conditions, at the same time providing a tool for the high-precision computer-controlled generation of patterns within a wide range of length scales from 1 μm down to 3 nm. By using highly linearized positioning components (3.5 nm repositioning accuracy / PI Karlsruhe, Germany) a range of up to 800 x 800 μm^2 can be achieved. In a further combination of AFM structuring and a subsequent chemical reaction, protruding polymeric line patterns can be generated as a new way of constructive nanolithography [7].

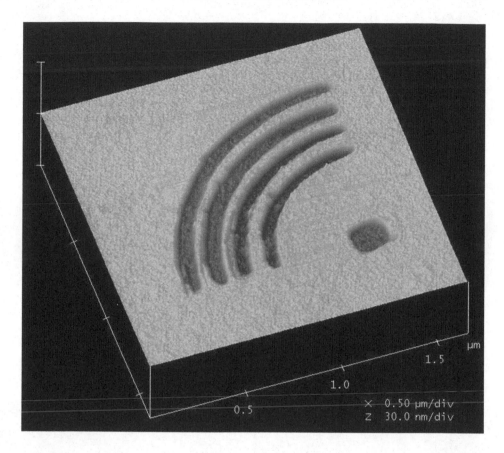

Fig. 3. Three dimensional image of another nano-machined structure in mica.(Depth: 3-5 nm)

Acknowledgments. We thank Patrik Dupeyrat for his experimental support.

References

1. Schmidt-Hansberg, B., Klein, M.F.G., Peters, K., Buss, F., Pfeifer, J., Walheim, S., Cols-mann, A., Lemmer, U., Scharfer, P., Schabel, W.: In-situ monitoring the drying kinetics of knife coated polymer-fullerene films for organic solar cells. APL (2009) (accepted for publication)
2. Böltau, M., Walheim, S., Mlynek, J., Krausch, G., Steiner, U.: Surface-Induced structure formation of polymer blends on patterned substrates. Nature 391, 877 (1998)
3. Geldhauser, T., Boneberg, J., Leiderer, P., Walheim, S., Schimmel, T.: Generation of surface energy patterns by pulsed laser interference on self-assembled monolayers. Langmuir 24(22), 13155–13160 (2008)
4. Geldhauser, T., Walheim, S., Schimmel, T., Leiderer, P., Boneberg, J.: Influence of the Relative Humidity on the Demixing of Polymer Blends on Prepatterned Substrates Macromolecules (accepted, 2009), doi:10.1021/ma9022058

5. Walheim, S., Schäffer, E., Mlynek, J., Steiner, U.: Nanophase-Separated Polymer Films as High-Performance Antireflection Coatings. Science 283, 520 (1999)
6. Müller, M., Fiedler, T., Gröger, R., Koch, T., Walheim, S., Obermair, C., Schimmel, T.: Controlled Structuring of Mica Surfaces with the Tip of an AFM by Mechanically Induced Local Etching. Surface and Interface Analysis 36, 189–192 (2004)
7. Barczewski, M., Walheim, S., Heiler, T., Błaszczyk, A., Mayor, M., Schimmel, T.: High Aspect Ratio Constructive Nanolithography with a Photo-Dimerizable Molecule. Langmuir (2009) (accepted for publication)

Challenges of Precision Assembly with a Miniaturized Robot

Arne Burisch, Annika Raatz, and Jürgen Hesselbach

Technische Universität Braunschweig,
Institute of Machine Tools and Production Technology
Langer Kamp 19 b, 38106 Braunschweig, Germany
a.burisch@tu-bs.de

Abstract. The first part of the paper describes the miniaturized robot Parvus, which is suitable for desktop factory applications. The Parvus is well equipped for pick-and-place of micro parts in precision assembly. The challenges of precision assembly are discussed considering the technical data and behavior of the robot. The hybrid robot operates based on parallel kinematics and is driven by micro harmonic drive gears. Due to its size-reduction, the Parvus offers prospects, but also constraints which are discussed and presented by measuring data. Finally, solutions for improving the precision of the robot are presented.

Keywords: Desktop factory, miniaturized robot, micro gripper, micro gear.

1 Introduction

Nowadays a trend of miniaturization with regard to product development can be observed in several industrial sectors. The Nexus III market study predicts that the market of millimeter-sized MST-products (Micro System Technology) will grow 16 % per year. However, an increasing gap can be observed concerning the dimensions and costs between the products and the production systems used. Assembly lines and clean rooms for millimeter-sized products often measure some tens of meters and are mostly too expensive for small- and medium-sized businesses. Thus, many micro-products are assembled by hand, which results in high assembly costs that amount to 20 % to 80 % of the total production costs [1]. A solution to prevent this could be the cost-optimized, flexible desktop factory for micro production.

2 The Miniaturized Robot Parvus

The robot Parvus, as shown in Figure 1, is a size-adapted handling device using innovative, miniaturized machine parts. With size-adapted handling devices, in the range of several centimeters to a few decimeters, easily scalable and highly flexible production technology can be achieved. The challenge of the functional model Parvus was to develop a miniaturized precision industrial robot with the full functional range of larger models.

S. Ratchev (Ed.): IPAS 2010, IFIP AICT 315, pp. 227–234, 2010.

The robot consists of a typical parallel structure, driven by Micro Harmonic Drive [2] gears combined with Maxon electrical motors. This plane parallel structure offers two translational DOF in the x-y-plane. The z-axis is integrated as a serial axis in the base frame of the robot. The easy handling of the whole plane parallel structure driven in z-direction is possible due to the minimized drive components and light aluminum alloy structure. The rotational hand axis Ψ was designed as a hollow rotational axis to be the Tool Center Point (TCP) of the parallel structure. This allows media such as a vacuum to be passed along the hand axis. This axis, with a diameter of 2.5 mm, can be equipped with several types of vacuum grippers. The development of the Parvus, its fundamentals, the miniaturized drive systems, and the robot design approaches have already been described in several previous papers [3].

Fig. 1. Two functional models of the miniaturized robot *Parvus*

3 Challenges of Precision Assembly

In most cases, the precision assembly of MST-products needs a highly precise pick-and-place application of the related parts. One example is the placement of optical ball lenses in a micro-optical LIGA bench with accuracy in the range of 1 μm [4]. This can be achieved by passive alignment and a pick-and-place process using a vacuum gripper. Other parts, such as prisms and photodiodes, can also be mounted in this way. The demonstrators of micro-electric actuators developed within the collaborative research center 516 in Braunschweig and Hannover also need a pick-and-place application of small ruby balls with a diameter of a few hundred micrometers [5].

For these pick-and-place applications, it is essential that the device always reaches the same assembly position multiple times. This is well described and characterized by the repeatability of the handling device following the standard of EN ISO 9283 [6].

The repeatability can be taken as a reference value for the maximum position accuracy of the robot end-effector. It is hereby assumed that the robot always reaches the position coming from the same direction.

However, in the case of picking several different parts from magazine trays around the assembly area, the handling device will reach the position from different directions. To get further information about the behavior of the robot in this case, the multiple direction position accuracy [6] has to be taken into account.

Another exemplary scenario is picking a glass ball out of a channel and moving the end-effector precisely along this channel. In this case, good path accuracy [6] of the robot is necessary to move a gripper precisely inside a channel without collision.

4 Prospects and Constraints of the First Functional Model Parvus

In most cases of typical pick-and-place applications, the Parvus is well-suited for a high precision assembly process. Figure 2 shows the robot being equipped with a vacuum gripper and a pneumatic mechanical micro gripper [7] for gripping a variety of different micro parts. The repeatability of the robot end-effector was measured within the primary (mainly used) workspace, as shown in Figure 3. The worst value of the repeatability is 5.7 µm (at measuring point P3, with 3 Sigma). This measured value is in the range of the expected repeatability of 2 – 6 µm, which was simulated within the primary workspace and based on an angular repeatability of 0.0027° [2] of the micro gears. As described in previous papers, this value was achieved by optimizing the stiffness of the micro gears.

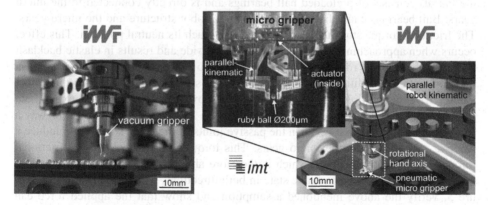

Fig. 2. Parvus with vacuum gripper (left), micro gripper (right)

As stiffness of the micro gears is important for the repeatability of the robot, it is also very important for the multiple direction accuracy. Stiffness related factors, including backlash and hysteresis in the robot drives, have a significant influence on the results. The multiple direction accuracy has been measured within the primary workspace at points P1, P2 and P5. The results show that these values differ 10 - 30 times as much as the repeatability.

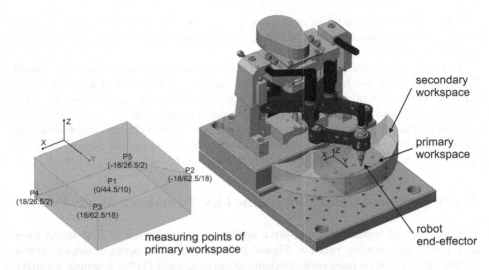

Fig. 3. Workspace and measuring points of Parvus

It has been investigated why there is such a difference between the multiple direction accuracy of the robot and the repeatability, even though the micro gears are backlash free. In this case, the relatively low stiffness of 2.6 - 6.13 Nm/rad [8, 2] of the micro drives compared to the dimensions of the robot structure causes this effect. When unloaded, the micro gears are free of backlash. However, the robot structure of the Parvus consists of preloaded ball bearings and is directly connected to the micro gears. Ball bearings cause frictional torque in the robot structure and the micro gears. The frictional torque stops the gear before it can reach its neutral position. This effect occurs when approaching from the left or the right side and results in elastic backlash in the combination of robot structure with the micro gears. If there were any backlash in the system, an additional pre-load in the robot structure would reduce this effect and improve the multiple direction accuracy. To prove this assumption, an experiment with preloaded robot structure was carried out, as shown in Figure 4. There was a spring force F_{spring} applied between the passive joints of the robot structure to induce a reactive torque M_{gear} in the micro gears. This torque was higher than the frictional torques of the robot structure, which is therefore able to deflect the gear beyond its neutral position into a more stable state in both directions. The results, shown in Figure 5, verify the above mentioned assumption and show that the applied force can improve the multiple direction accuracy by up to 3 times, here from 55 μm to 18 μm. The experiment with the spring is actually possible in a very small area of the workspace. The small range of applicable spring forces between the robot joints, limits the movement of the robot structure. Therefore, different springs have to be used for different points in the workspace. To implement a stable preload on the robot drives without limiting the workspace, the torque has to be applied directly to the drives. Here a torsion spring or an active torque generator has to be connected to the output of the micro gears.

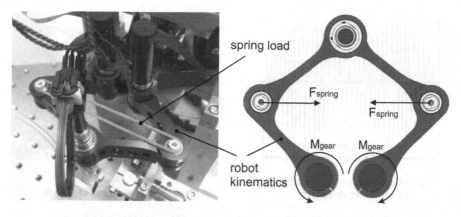

Fig. 4. Robot kinematics with spring load and reaction torque

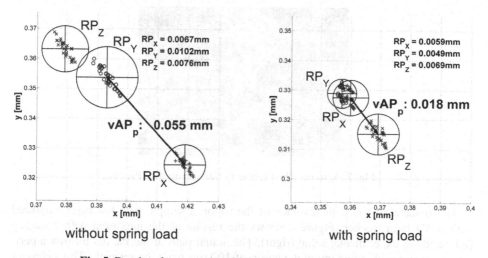

Fig. 5. Results of measured multiple direction accuracy at point P5

In addition, there are also other influences of the micro gears concerning the characteristics of the robot kinematics. In general, harmonic drive gears show special behavior in kinematic error and transmission compliance, as shown in Figure 6 (left) [9]. At the high grade of miniaturization of the micro harmonic drive gears used in the robot, theses effects immensely influence the path accuracy of the entire robot, as shown in Figure 6 (right). In most cases, the path accuracy is only important for machine tools and not for pick-and-place robots. However, in some cases it has to be considered that there is a deviation of the set path.

One example is the automatic ball feeder (designed at IMT of TU Braunschweig), where the Parvus has to move its micro gripper inside a channel for gripping a pre-positioned glass ball, as shown in Figure 7. Here the gripper has to move along a relatively straight line, which requires good path accuracy.

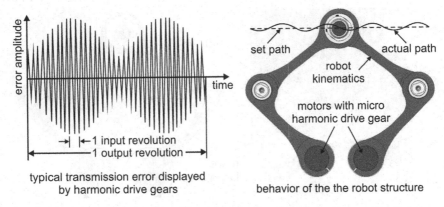

Fig. 6. Transmission error of harmonic drive gears [9] and behavior of the Parvus

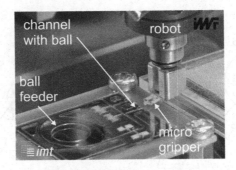

Fig. 7. Automatic ball feeder in interaction with Parvus

To characterize the path accuracy of the robot, a sample path has been measured with a 3D laser tracker. Figure 8 shows the results of the measured path accuracy (left) and the experimental setup (right). The actual path of the Parvus follows a periodic behavior with a maximum deviation of 100 µm from the set path. This behavior is obviously caused by the transmission error of the micro gears. The transmission error inside the gear can also be described as a kinematic effect. There cannot occur any dynamic effects, thus the measurements have been done at very low speed.

In order to optimize the path accuracy of the robot, it is necessary to measure the transmission error at the output of each gear. This is necessary because the path accuracy of the robot shows a combination of the transmission errors of both drives of the parallel robot structure. The best way to detect this behavior is using a small and precise angular sensor. If it is possible to obtain precise information about the driving angle during operation, the control could compensate the transmission error of the gears and help improving the path accuracy of the robot. A small precise angular sensor is currently under testing. Another approach is the use of computed feed forward control to reduce disturbing effects coming from the gears. This is under examination using the inverse dynamic model of the Parvus.

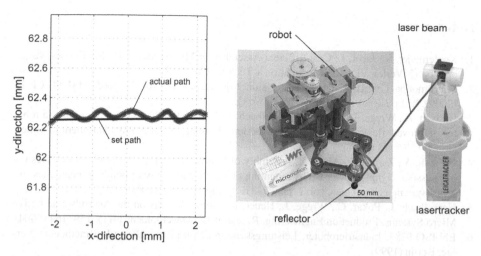

Fig. 8. Measured path accuracy of the Parvus (left), measuring setup (right)

Furthermore, the above mentioned measurements (repeatability, multiple direction accuracy and path accuracy) of the robot kinematics driven by micro harmonic drive gears will also be done with the same robot kinematics connected to small harmonic drive gears RSF-5A. This can help to ensure that there are no disturbing effects as a result of the kinematics and ball bearings.

5   Conclusions and Outlook

The experiments and measurements with the Parvus have shown that the robot is well suitable for pick-and-place applications with repeatability exceeding 5 μm. It has been shown that the stiffness of the micro gears influences the repeatability of the robot. The multiple direction accuracy of the actual functional model still has to be improved. The experiment has demonstrated that preloaded torque can improve this behavior. To ensure this, further experiments have to be performed. A way must be found to implement a stable preload on the robot drives without limiting the workspace. It has been observed that there is a deviation from the set path of the robot. Therefore, path accuracy will be measured and improved. The required hardware and control strategies are currently under testing. The results of measurements and strategies for improving the accuracy of the robot will be used for the design of the next prototype of Parvus.

Acknowledgement

The authors gratefully acknowledge the funding of the reported work by the German Research Center (Collaborative Research Center 516 "Design and Manufacturing of Active Micro Systems").

References

1. Koelemeijer, S., Jacot, J.: Cost Efficient Assembly of Microsystems. In: MST-news, vol. 1, pp. 30–32. Verlag VDI/VDE Innovation + Technik GmbH (1999)
2. Micromotion GmbH, Catalogue: Precision Microactuators, vol. 04/2005 MM 90 01 18 (2009), http://www.micromotion-gmbh.de
3. Burisch, A., Wrege, J., Raatz, A., Hesselbach, J., Degen, R.: PARVUS – miniaturised robot for improved flexibility in micro production. Journal of Assembly Automation, Emerald 27(1), 65–73 (2007)
4. Gerlach, A., Ziegler, P., Mohr, J.: Assembly of hybrid integrated micro-optical modules using passive alignment with LIGA mounting elements and adhesive bonding techniques. In: Microsystem technologies, vol. 7, pp. 27–31. Springer, Berlin (2001)
5. Hesselbach, J., Pokar, G., Wrege, J., Heuer, K.: Some Aspects on the Assembly of Active Micro Systems. Production Engineering. Research and Development 11(1), 159–164 (2004)
6. EN ISO 9283: Industrieroboter: Leistungskenngrößen und zugehörige Prüfmethoden. Verlag, Berlin (1999)
7. Hoxhold, B., Büttgenbach, S.: Micro Tools with Pneumatic Actuators for Desktop Factories. Sensors and Transducers 7(10/09), 160–169 (2009)
8. Slatter, R., Degen, R.: Miniature Zero-Backlash Gears And Actuators For Precision Positioning Applications. In: Proc. of 11th European Space Mechanisms and Tribology Symposium ESMATS 2005, Lucerne, Switzerland, September 21-23, pp. S.9–S.16 (2005)
9. Tuttle, T.D.: Understanding and Modeling the Behavior of a Harmonic Drive Gear Transmission. Technical Report: AITR-1365, Massachusetts Institute of Technology, Cambridge, MA, USA (1992)

PART IV

Development of Micro-assembly Production Systems

Chapter 8

Modular Reconfigurable Assembly Systems

Application of a Reconfiguration Methodology for Multiple Assembly System Reconfigurations

D. Smale and S. Ratchev

Precision Manufacturing Centre, Faculty of Engineering, University of Nottingham,
Nottingham, UK

Abstract. Reconfigurable Assembly Systems (RAS) offer the potential to enable rapid exchange of functional modules. There has however been little investigation into the planning of multiple system reconfigurations. The work proposes a capability-based approach; consisting of a Reconfiguration Methodology, supported by a Capability Model and Taxonomy. The Methodology focuses on multiple system reconfigurations and is based upon operator-oriented definition, thereby utilising existing knowledge and expertise. The Capability Model consists of Capability Identification and Comparison processes: by aggregating the results, capability and compatibility sets can be derived. Further, the Model has strong links to the Capability Taxonomy. The work describes the overall approach and key elements and provides a detailed example application in software using a simple multi-product test case. Key conclusions are drawn and future work outlined.

Keywords: Reconfigurable Assembly, System Planning, Capabilities.

1 Introduction

Reconfigurable Assembly Systems (RAS) offer the ability to rapidly exchange process equipment modules, which facilitates a change in product, system and/or to provide equipment redundancy. There have been efforts in developing platforms that can be physically reconfigured [1] however; there has been little investigation into the planning of multiple system reconfigurations.

The current state-of-the-art in manufacturing systems dictates that the vast majority of new lines are bespoke and single purpose. Generally, the system is designed to be as cost effective as possible [2]. RASs are researched as a means of addressing the key issues presented by conventional systems [3]. The reconfigurability is typically achieved through the implementation of standardised mechanical, electrical and control interfaces. RASs are widely accepted as the route for future production systems [4]. There has been some effort towards identification of requirements for new or reconfigured assembly systems [5], the design of new assembly systems [3] and operation allocation [6]. However, a methodology which considers multiple system reconfigurations, and which is applicable both to new and to existing systems, is not available.

S. Ratchev (Ed.): IPAS 2010, IFIP AICT 315, pp. 239–246, 2010.
© IFIP International Federation for Information Processing 2010

2 Methodology Overview

In order to achieve the full potential offered by RAS, a new methodology must be developed. It is proposed that this includes three key elements: 1) Capability taxonomy and definition process, 2) Capability modeling and comparison and 3) Reconfiguration identification, optimization and validation methods. These elements can be further enhanced by focusing on precision production, micro-manufacturing and assembly rather than machining. The proposed approach is intended to broadly follow the, currently human-centred, complex decision-making approach from the specification of requirements by the customer through to the specification and planning of the assembly system solution. The proposed approach sequentially follows four key process elements: Capability Modeling, Capability Identification, Capability Comparison and Configuration and Sequence Analysis.

2.1 Capability Modeling

The modeling of the Capabilities fundamentally requires that there is a clear definition for "Capability". This definition must be applicable to both the products required and the available equipment. Furthermore, enabling a tiered or hierarchical definition with several levels of detail is of substantial benefit as this enables the application of the model to various stages in product development. This definition will constitute a Capability Taxonomy.

The proposed Capability Taxonomy is described in detail in earlier work [7]. It comprises 6 Capability Classes: *Motion, Join, Retain, Measure, Feed and Work.*

In addition to the clear Taxonomy, it is essential that the requirements for the project are defined. This in turn requires a clear structure of the project and the major stakeholders: a factor which is made more important when considering specification of the system early in the product development cycle. These matters are researched and solutions defined in previous work [8].

2.2 Capability Identification

The next stage is to identify each capability; both Required and Available. The identification of the Available Capabilities will be supported by guidelines associated with the Taxonomy. These guidelines provide a series of Yes/No questions which enable the capabilities provided by the equipment to be identified.

In order to identify the capabilities that are required by each product in order to produce them, a more structured method is required. This is termed the Process Flow Template (PFT) and is based upon the principle that assembly requires two parts to be brought and maintained together. Therefore, the core assembly processes are "Moving Part x" and "Joining Parts x and y". By using this as a basis for the definition of the required capabilities, a number of rules can be applied. The application of these rules to the Process Flow Template diagram produces a Capability Flow Diagram (CFD). This diagram identifies the required capabilities for one product: therefore, one diagram is needed for each product. The Capability Taxonomy is then applied in order to define each individual capability. The full details of the identification process are described in previous work [7].

2.3 Capability Comparison

The next step is to perform a comparison of the capability sets. For this purpose a Comparison Matrix is used. This concept is developed and an example provided in earlier work [7]. The matrix has been devised in order to enable two or more capability sets to be compared. Existing capabilities are listed in the first column, whilst required capabilities are listed across the first row. At each intersection a comparison is performed. Then, each row and column is totalled in order to provide the basis for configuration analysis. The result of the comparison will be one of three values: '1' meaning that the capabilities match, '0.1' meaning the capabilities are compatible but not exactly matching and '0' meaning that there is no realisable commonality between the capabilities. The '0.1' value can be viewed as the 'grey area' of the analysis and is used to represent cases where the existing capability is over-specified with respect to the required capability. Whether or not the existing capability is used or replaced is a decision based more upon the requirements of the system as a whole rather than purely on the technical matters.

2.4 Configuration and Production Sequence Analysis

The next stage of the methodology is to analyse the comparison results to determine the optimum configuration for each product. This analysis will focus on the previous determined 'grey area' of capabilities: those with a 0.1 value in their column of the Comparison Matrix. The decisions made and the methodology which supports them, has been described and demonstrated in earlier work [9].

At this point in the analysis, each product has an optimized solution configuration. However, there are still a number of decisions to be made and issues to be resolved, specifically finding the Production Sequence. This is the order in which each of the products is produced and is one of the most critical decisions affecting the performance and efficiency of the system. It is generally preferable to minimize the system disruption, downtime and reconfiguration effort (module exchange), thereby minimizing the overall cost of reconfiguration. This is achieved through adopting a 'product-centered' approach and identifying the commonalities between each of the products. The Production Sequencing Method is detailed below:

A new matrix is used to provide this analysis; the correlation matrix enables comparison between the Required Capabilities for the different products. As with the previous comparison process, a '1' is entered if the Capabilities are compatible, '0' is entered for non-matches. The details of the exact calculation of the Product Correlation Ratio (PCR) is shown and demonstrated in [9].

3 Example Application

To demonstrate the process, a highly simplified example is used. The example used is based on real equipment and processes, but considers only a very short list so that it is easy to follow and concise for the purpose of this research. One of the key innovations of this research is the consideration of multiple reconfigurations thus the example contains five products but each with only a few processes for conciseness.

Define the Existing System Capabilities. The example existing system consists of: *one SCARA type robot, one mechanical gripper, one tray feeder and a static fixture.*

At this stage the Capability Taxonomy and the associated guidelines are used to define the capabilities for the existing equipment. Most equipment modules have only one capability; more flexible modules can have multiple capabilities which makes the definition more complex but in this example, each module has one capability:

- **Robot:**   Motion 1,1,39,2,2,3
- **Gripper:**   Retain 3,1,2,2,
- **Tray:**   Feed   2,3,2,2,
- **Fixture:**   Retain 3,2,2,2,

Define the Required Capabilities. The products used for this example are:

- Product A: Cap onto a Cylinder.
- Product B: Chip on a PCB.
- Product C: Pin on a Plate.
- Product D: Sphere onto a Shaft.
- Product E: Cube into a Slot.

Using the PFT five CFDs are produced, one for each product. This enables the identification of all of the possible Required Capabilities for each product. An example of this, for Product B, is shown in Fig. 1. The five capability sets can be listed and defined and are shown in Table 1.

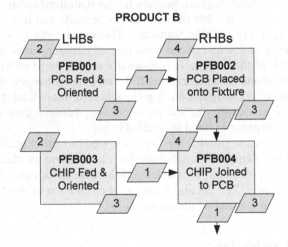

Fig. 1. The completed CapID Diagrams for the 5 products

Perform Capability Comparison. The first task is to sequence the capabilities. This is important for the selection and optimisation procedures later. Additionally, the capabilities are given a short alpha-numeric designator. Even a very simplified example of 5 2-part products results in a total of 55 required capabilities; for this reason only those capabilities for Product B (along with their definition from the Taxonomy) are shown in Table 2 below. This clarifies the need for the process to be automated within software. The sequenced capabilities can then be entered into the Capability Matrix. The first stage of analysis investigates each configuration separately. Each Required Capability Set is compared to the Existing Capability Set with the result being entered into the intersecting location in the matrix. The Capability Matrix for Product B is illustrated in Table 3.

Table 1. List of all of the capability sets by product

Product A	Product B	Product C	Product D	Product E
PFA01	PFB01	PFC01	PFD01	PFE01
PFA01	PFB01	PFC01	PFD01	PFE01
PFA01-PFA02	PFB01-PFB02	PFC01-PFC02	PFD01-PFD02	PFE01-PFE02
PFA02	PFB02	PFC02	PFD02	PFE02
PFA02	PFB02	PFC02	PFD02	PFE02
PFA02-PFA04	PFB02-PFB04	PFC02-PFC04	PFD02-PFD04	PFE02-PFE04
PFA03	PFB03	PFC03	PFD03	PFE03
PFA03	PFB03	PFC03	PFD03	PFE03
PFA03-PFA04	PFB03-PFB04	PFC03-PFC04	PFD03 PFD04	PFE03-PFE04
PFA04	PFB04	PFC04	PFD04	PFE04
PFA04	PFB04	PFC04	PFD04	PFE04
PFA04-OUT	PFB04-OUT	PFC04-OUT	PFD04-OUT	PFE04-OUT

Table 2. The full definition of the capabilities for Product B

Cap. Locator	Cap. Designator	Cap. Definition
PFB01-PFB02	PB01	1,1,39,2,2
PFB03-PFB04	PB02	1,1,39,2,2
PFB02-PFB04	PB03	1,2,-
PFB04-OUT	PB04	1,2,-
PFB01	PB05	2,2,1,1,1
PFB03	PB06	2,2,2,1,1
PFB01	PB07	3,1,1,1,1
PFB03	PB08	3,1,2,1,1
PFB02	PB09	3,2,1,1,1
PFB02	PB10	4,1,2,2,1,4
PFB04	PB11	4,1,2,2,5,4

From the results of the five comparison matrices for products a to E, the following conclusions can be drawn with respect to the original equipment capabilities:

- The existing robot can be used in production of Products A, B, C and E
- The existing Tray Feeder cannot be used in production of any products
- The existing Mechanical Gripper can be used in production of Product B and E and possibly in the production of Products B and C

Table 3. The Capability Matrix for Product B

		REQUIRED CAPS										Total C_{EXS}
		1. Motion				2. Feed		3. Retain			4. Join		
		PB01	PB02	PB03	PB04	PB05	PB06	PB07	PB08	PB09	PB10	PB11	
EXISTING CAPS	Ex01	1	1	0	0	-	-	-	-	-	-	-	2
	Ex02	-	-	-	-	0	0	-	-	-	-	-	0
	Ex03	-	-	-	-	-	-	:1	1	0	-	-	1:1
	Ex04	-	-	-	-	-	-	0	0	:1	-	-	:1
Total C_{REQ}		1	1	0	0	0	0	:1	1	:1	0	0	

- The existing Fixture can be used I the production of Product E and possibly in the production of Products B and C

Perform Configuration Analysis. The first element of the configuration analysis is to analyse the configuration for each product independently. The overall analysis of the complete system, including equipment allocation, is the role of the Reconfiguration Methodology. At this stage of analysis, the capabilities are divided into four lists: Retained, Redundant, Investigation and Procurement. A summary of this evaluation for Product B is illustrated in Table 4.

- Retained Capabilities are the existing capabilities with a value total of 1 or more. These capabilities should be re-used in the new configurations.
- Redundant Capabilities are the existing capabilities with a value total of 0. These capabilities will be removed and not used in the new configurations.
- Investigation Capabilities are the existing capabilities with a value total of :1 or more. These could be used in the new configurations, but their applicability will depend upon further analysis which will be conducted in the Reconfiguration Methodology.
- Procurement Capabilities are the Required Capabilities that are not met by any of the existing capabilities. These must be procured.

Table 4. A summary of the capability evaluation for Product B

Product B	
Retained Capabilities	Ex01, Ex03
Redundant Capabilities	Ex02
Investigation Capabilities	Ex03, Ex04
Procurement Capabilities	PB03-PB06, PB10, PB11

Perform Sequence Analysis. The first element of the sequence analysis is to construct one amalgamated comparison matrix. All of the five capability sets are listed in a single matrix, this matrix is then extended vertically with a triangular "House of

Quality" style inter-relationship grid. This is used to perform the same kind of comparison of capabilities as performed previously but between the different required sets. With each capability compared, a ratio is derived to determine the similarity between the requirements (and hence the likely similarity between the resulting configurations). This is termed the Similarity Coefficient and is defined in the form:

$$\textbf{\textit{Similarity Coefficient}} = \frac{\textit{\{No. of matching capabilities\}}}{\textit{\{No. of possible matching capabilities\}}}$$

In this case, the matching capabilities are those with the exact match. The number of possible matching capabilities is defined as the number where a comparison is made (different Classes are not compared and so not included in the total). The full matrix is not shown due to its size. Table 5 shows the similarity comparison results.

Table 5. The Similarity Coefficients for all of the five products in the example.

	PROD A	PROD B	PROD C	PROD D	PROD E
PROD A	N/A	6/33	7/33	8/33	5/33
PROD B	Repeat	N/A	4/33	5/33	17/33
PROD C	Repeat	Repeat	N/A	2/33	13/33
PROD D	Repeat	Repeat	Repeat	N/A	6/33
PROD E	Repeat	Repeat	Repeat	Repeat	N/A

This table enables the optimal production sequence to be indentified. This sequence is the one in which the system downtime is minimized, which has a direct correlation to the number of capability exchanges required to deliver the new configuration. Thus the sequence is identified by finding the production sequence which has i) the highest total similarity value and ii) a feasible sequence delivering all of the products. In this example the four highest value relationships are: BE, CE, AD and AC. This results in two possible sequences:

1. B→E→C→A→D
2. D→A→C→E→B

As one is the reverse of the other, the decision as to which one to choose is based upon the similarity of the two products at the ends of the sequence to the existing system. The result of this is that the final production sequence is:

B→E→C→A→D.

4   Conclusions and Further Work

This paper has presented a methodology for the planning of multiple reconfigurations of a reconfigurable assembly system. This methodology has been demonstrated through a simple example and shown to aid in the identification of capabilities to remove, retain and procure. Furthermore it has demonstrated that the optimal production sequence can be identified.

The next stages of work will address the finer details of the Reconfiguration Methodology and it's application to software. It will also investigate the equipment allocation, which will include developing a library of equipment modules. Furthermore, more detailed test cases based upon commercially available/destined products will be developed to apply to and assess the methodology.

Acknowledgments

The authors would like to acknowledge the Engineering and Physical Sciences Research Council (EPSRC) in the United Kingdom for the funding which has enabled this research to be conducted.

References

1. Koren, Y.: Report to Industry. The NSF Engineering Research Center for Reconfigurable Manufacturing Systems, University of Michigan College of Engineering (2004)
2. Du, S., Xi, L., Ni, J., Ershun, P., Liu, C.R.: Product lifecycle-oriented quality and productivity improvement based on stream of variation methodology. Computers in Industry 59(2-3), 180–192 (2008)
3. Vos, J.A.W.M.: Module and System Design in Flexibly Automated Assembly, TU Delft. PhD (2001)
4. Lohse, N.: Towards an Ontology Framework for the Integrated Design of Modular Assembly Systems, The University of Nottingham. PhD (2006)
5. De Lit, P., Delchambre, A.: Integrated Design of a Product Family and its Assembly System. Kluwer Academic Publishers Group, The Netherlands (2003)
6. Sujono, S., Lashkari, R.S.: A multi-objective model of operation allocation and material handling system selection in FMS design. International Journal of Production Economics 105(1), 116–133 (2007)
7. Smale, D., Ratchev, S.: A Capability Model and Taxonomy for Multiple Assembly System Reconfigurations. In: 13th IFAC Symposium on Information Control Problems in Manufacturing, Moscow, Russia, June 3-5. Elsevier, Amsterdam (2009)
8. Smale, D., Ratchev, S.: Enabling a Reconfiguration Methodology for Multiple Assembly System Reconfigurations. In: COMEH (Consortium of UK University Manufacturing and Engineering) ICMR International Conference on Manufacturing Research, Warwick, United Kingdom, September 8-10 (2009)
9. Smale, D., Ratchev, S.: A Reconfiguration Methodology for Multiple Assembly System Reconfigurations. In: ASME (American Society of Mechanical Engineers) 4th International Manufacturing Science and Engineering Conference, West Lafayette, IN, USA, October 4-7 (2009)

Multi-agent Architecture for Reconfiguration of Precision Modular Assembly Systems

Pedro Ferreira, Niels Lohse, and Svetan Ratchev

The University of Nottingham, Precision Manufacturing Centre, Nottingham,
NG7 2RD, United Kingdom
Tel.: +44 115 951 3875
epxpf@nottingham.ac.uk

Abstract. Precision assembly systems today are subject to high levels of change which require fast physical and logical adaption. As such, new methods and tools need to reflect this need providing adaptive systems that can react to market changes in a timely and cost effective manner. Towards achieving increased responsiveness several developments of new modular concepts provided the bases for higher system adaptability through increased module interchangeability and reusability. The modularization of physical and control infrastructure only solves one aspect of the issue, there is still a lack of appropriate tools and methods to support requirements driven reconfiguration of such systems. This paper proposes an agent architecture for reconfiguration of equipment modules driven by a set of requirements. A new agent model is proposed which addresses the specific needs of precision modular assembly systems catering both for physical and logical constraints of the modules as well as their joint emergent behaviour.

Keywords: Reconfigurable Assembly Systems, Modular Assembly Systems.

1 Introduction

The question of component reusability and dynamic reconfiguration of precision assembly systems has become increasingly more important due to ever decreasing product life-cycles, increasing product variants and rising process complexity. General purpose assembly machines, equivalent to CNC machine tools, are only available in specialist domains such as printed circuit board assembly where the components are highly standardised. The assembly of most other products demand custom made systems which address the specific requirements for these products. Today, these are mostly "Engineered to Order" making them cost and time intensive to design or reengineer. Reconfiguration of assembly systems requires a certain level of flexibility. This can be achieved by adding more equipment to the system, however adding equipment for future use can be very costly. Modular systems offer the structure to add or remove equipment as and when it is needed, thus enabling system capabilities to be continuously adapted to changing requirements over the lifecycle of the system. Increased modularisation of assembly equipment, rapid integration mechanisms, and

S. Ratchev (Ed.): IPAS 2010, IFIP AICT 315, pp. 247–254, 2010.
© IFIP International Federation for Information Processing 2010

automatic configuration tools are considered fundamental for the move towards cost and time effective configuration and re-configuration of complex assembly systems [1, 2, 3] .

Significant effort has been directed towards creating modular assembly system (MAS) architectures for physical equipment and control interchange ability. The EU project EUPASS, for instance, has created a framework for rapid integration for ultra-precision assembly modules defining hardware interfaces, control interfaces, and module description formats [4]. Other modular assembly system platforms have been proposed [5, 6]. While the number and completeness of underlying industrial applicable standards is still limited, there is a clear drive to overcome this barrier.

Standardisation of hardware and software interfaces is, however, only one aspect of rapid assembly system reconfiguration. Effective tools and methods for the requirements driven system analysis, reconfiguration and validation of complex assembly system solutions are also needed to drastically reduce the time and effort required for the development of highly dedicated assembly systems. MAS reconfiguration methods reported today adopt a human driven approach, which is based on the expertise of the user. These methods provide valid solutions, however, very few of these are replicable and transparent, and are constrained by the individual knowledge of the user.

This paper proposes a multi-agent architecture for reconfiguration of precision MAS based on the changed product and process requirements. The proposed architecture is directed towards utilising latest agent-based negotiation protocols to enable the scalable and extendable requirements driven reconfiguration of complex MAS assessing the reconfiguration effort and providing solutions from the existing modules if possible. The system can also access a library of existing modular building blocks to enhance the system with the required functionality.

2 Literature Review

The term reconfiguration implies that something has been configured and requires a new configuration. The basic principles of a configuration still apply a reconfiguration, although the decision making process will differ quite substantially. The use of different methodologies for configuration design depends on the complexity of the system [7], as for the reconfiguration. Systems can be classified as an uncoupled system, loosely-coupled system, or strongly-coupled system. MAS are normally comprised of strongly-coupled and loosely-coupled systems. A reconfiguration methodology for MAS has to cater solutions for the presence of both.

Reconfiguration can be seen as an enhanced configuration problem, e.g. the addiction of extra requirements and constraints. Configuration design of MAS is comparable to the design of modular products [7]. Therefore there are many methods that can be applied such as feature-based methods [8], hierarchical decomposition methods [9], combinatorial synthesis method [10], entity-based methods [11], and case-based methods [12]. In strongly-coupled systems all design variables have to be considered together to validating if the configuration fulfils its requirements [7].

The configuration design at system level is usually achieved through a propose-and-test approach in conjunction with system simulation software for testing. An approximate solution is found in a time-consuming iterative process. Mathematical formulation for the system level would be too complex and it is usually only used for specific sub-problems [7]. Deterministic models where the system variables are constant have been used in configuration design [13], these do, however, limit the flexibility of the design system. Stochastic models arise as a solution to this problem since they provide at least one uncertain variable. Some configuration design methodologies have used stochastic models in order to deal with the configuration problems [14, 15].

The research on reconfiguration of assembly systems is mostly in the control aspects of the assembly systems producing reconfigurable software which is able to change the control of the assembly systems yet falling short on physical reconfiguration and system enhancement. These approaches are also not related with the systems design and requirements specifications, and are mostly human driven.

Despite the significant work in the area of configuration and reconfiguration methods, there is still a lack a systematic reconfiguration methodology that caters for different reconfigurable machines and is driven by design requirements [16]. Furthermore, most of the proposed methods in the field of manufacturing system reconfiguration have been focused on the machine level, while the systems have been designed largely intuitively [7].

Agent technology is seen as the natural way to address the problems of scalability and flexibility manufacturing systems. As a result agent technology has been widely applied to provide solutions in the manufacturing domain [17]. This provides extensive literature in agent models, negotiation models, agent environments, etc, however these models are mostly application specific.

The exiting agent-based approaches are mainly focussed on providing agility and low level reconfigurability. For reconfiguration design, they also need to have the tendency to provide close to optimal solutions. The optimization in distributed systems is quite different from other approaches which target global optimization through mathematical formulation of the whole problems. The mathematical formulation of complex system is quite difficult to develop and maintain, thus are mainly successful for simpler systems. Agent approaches on the other hand attempt to achieve optimization through efficient coordination mechanisms and thus require significantly simpler models [17]. Despite all the work done in this field there is still no agent architecture which focuses on the specific issues of modular assembly systems identified by [18], namely providing clear formalisms for equipment capability and interconnection constraint representation as well as methods and protocols for system formation from these modules based on a given set of requirements.

3 Problem Definition

The redesign of assembly systems today is a largely human driven process relaying heavily on the skill and expertise of system integrators. This process can provide valid system reconfigurations, however, it seldom follows a systematic approach and as a result will often be quite expensive and time consuming, making frequent system adaptations infeasible. Furthermore, there is a lack of clear assessment methods and tools which can establish when a reconfiguration should occur.

The MAS paradigm with its focus on clear functional decoupling of equipment module functionalities and standardised interfaces for interchange ability has opened the scope for automatic reconfiguration methods. Thus, it is possible to clearly formalise the functional capabilities and connectivity constraints of the available modules hence allowing the mapping of required against available capabilities. The redesign of MAS is essentially a restructure of equipment and assembly process, across several levels of granularity with equipment modules and their functional capabilities as the elementary building blocks.

The MAS reconfiguration problem is centre on two questions "when to reconfigure?" and "how to reconfigure?". Current solutions rely on the system integrator to assess the change in the system requirements and cater for the new requirements, reconfiguring the system accordingly while making decision about the added benefit resulting from reconfigurations.

A reconfiguration methodology for MAS has to be able to assess and recommend whether an existing system should be maintained in its current configuration or should be adapted in some form to provide the best cost benefit for a new set of requirements. Within a MAS paradigm, the reconfiguration methodology should start with the existing system and the changes in the requirements, and base all its recommendations on the available set of equipment modules as illustrated in Fig. 1.

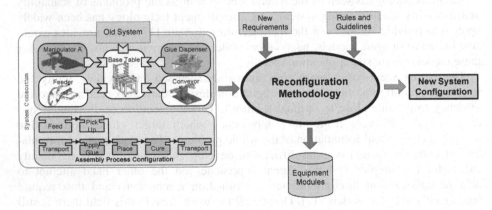

Fig. 1. Problem definition overview

4   Agent Architecture

This section gives a detailed overview of the proposed agent-architecture for MAS reconfiguration and explains the rational behind it. The GAIA methodology developed by [19] has been applied to translate the requirements identified for MAS reconfiguration problems above into an appropriate agent architecture. A schematic overview of the resulting agent architecture is given in Fig. 2.

The clear common denominator of all reconfiguration design methodologies including for MAS is the need to elicit and maintain the system requirements independent of the proposed solution alternatives. Consequently there is a need for Requirements Agents which are able to provide clear objectives to those agents involved in the reconfiguration process. Furthermore, they need to be able to represent the interests of the customer/system user to validate possible system reconfigurations.

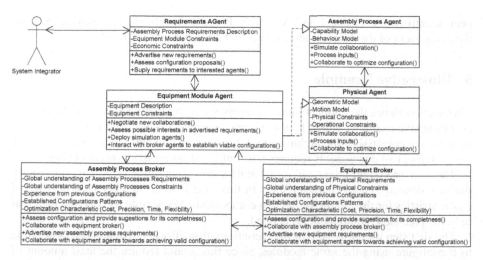

Fig. 2. Agent architecture for reconfigurable assembly systems

Another important set of actors within this problem domain are the equipment modules. It is proposed that each equipment module should be represented by an Equipment Module Agent that has a detailed understanding of the module's capabilities and behaviour. The role of mediator has been assigned to two types of agents as the configuration of MAS always needs to address both the equipment and the process: the Physical Broker Agent, and the Assembly Process Broker agents. Both are specialised to validate the logical constraints of physical and assembly process configurations respectively. These agents will negotiate tradeoffs between themselves in order to balance the equipment reconfiguration and assembly process reconfiguration constraints. This architecture requires some early evaluation of the likely success of reconfiguration within the existing consortium. Thus some method for early evaluation of the likely success the reconfiguration within the consortium needs to be available to reduce the computation effort. To provide some bases for early comparison, it is proposed that the Equipment Module Agents deploys Physical and Assembly Process Agents into a simulation environment. They represent the physical and process capabilities of the modules and dynamically interact with each other to determine the emergent behaviour and performance characteristics of a consortium. This will also provide valuable input for the benefits vs. effort analysis to assess if the reconfiguration should occur.

The whole process of reconfiguration is triggered by the requirements agents. This agent approaches the Equipment Module Agents involved in the established consortium with new requirements. The next step in the reconfiguration process happens through the negotiation of the Equipment Module Agents involved in the consortium and the Broker Agents that provide the expert knowledge for the reconfiguration. If the consortium is unable to provide a new solution, the broker agents will provide expert knowledge to enable phase two, the contact of other Equipment Module Agents that can enhance the consortium towards fulfilling the new requirements. This is followed by the negotiation between new and old equipment module agents and brokers to establish possible system alternatives. Once these system alternatives have been identified the Physical and Assembly Process Agents are deployed to provide the assessment of the different alternatives and determine their validity.

5 Illustrative Example

This section shows the application of the proposed agent architecture with the help of an illustrative example, which is both simple enough to follow and complex enough to show the potential of the architecture. Let us assume an automatic workstation for the gluing of two components requires a reconfiguration for a new product (Fig. 1).

The user or system integrator would start a new Requirements Agent and specify the new desired system characteristics. In this case it would simply be a new set of characteristics for the joining of the two components. For simplicity it is assumed that only the precision requirement is changed.

The first step for the system in the reconfiguration effort is to assess if it is possible to reconfigure using the same modules, since this would require the least amount of change. In this situation this analysis determines that all the modules have the minimum precision required. However once the Equipment Module Agents deploy the Assembly Process Agents for the assembly process simulation, it is determined that the errors stack up above the requirements for precision.

The Broker Agent is then contacted to provide strategies to compensate for lack of precision, namely the increase of precision by replacing less precise modules, or add measuring capabilities to compensate for stacked up errors. Given that the Assembly Process Agent during their analysis of the existing system have already determined which module is the bottleneck, the Equipment Module Agents can then proceed to finding alternatives based on the Broker Agent suggested alternatives. For the given example the bottleneck assembly process is "Place" which is performed by "Manipulator A". As such, the consortium will either find a way to compensate for the lack of precision by enhancing the system with a measuring module, or by replacing this module.

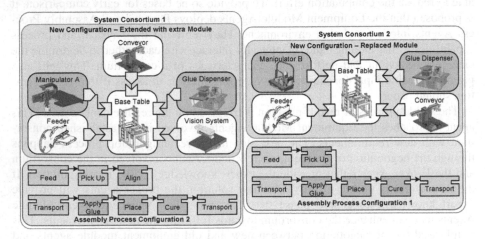

Fig. 3. Example of Alternative MAS Configurations

The Equipment Module Agents based on this knowledge advertise new module requirements for the system, which are in turn answered by relevant Equipment Module Agents outside of the original consortium. The possible alternatives are then evaluated based on their value to the consortium, and the most promising ones are simulated through the deployment of the Assembly Process Agents and Physical Agents. In this example the system would produce two promising reconfigurations for the system; adding a vision system to compensate for error; or replacing the manipulator with a more accurate one (Fig. 3). Because in both situations the system requires both physical and logical changes, the effort value is based on cost. By considering the cost of new modules as similar, it is clear that solution 2 has the advantage since the old module could be resold or reused in another system.

6 Conclusion

The presented agent architecture demonstrates clear potential to deal with MAS reconfiguration. This is expected to reduce reconfiguration times by better supporting the reconfiguration of MAS. This architecture also supports the constant analysis of a MAS to determine the reconfiguration effort at any given time. It is foreseen that the reconfiguration effort can be used against predicted benefits to determine the best time for reconfiguration.

Further work will focus on the development of agent negotiation and collaboration protocols which address the specific needs of MAS.

Acknowledgement

The reported work is partially funded by the European Union as part of the NMP-2-CT-2004-507978 EUPAS project. The support is gratefully acknowledged.

References

1. Koren, Y., et al.: Reconfigurable Manufacturing Systems. CIRP Annals - Manufacturing Technology 48(2) (1999)
2. Kratochvíl, M., Carson, C.: Growing Modular. Spinger, Berlin (2005)
3. Onori, M., et al.: European precision assembly roadmap 2012. The Assembly-Net Consotium (2002)
4. EUPASS. Evolvable Ultra-Precsion Assembly SystemS (2008),
   http://www.eupass-fp6.org/
5. Alsterman, H., Onori, M.: Process-Oriented Assembly System Concepts - The MarklV Approach. In: ISATP 2001, Fukuoka, Japan (2001)
6. Gaugel, T., Bengel, M., Malthan, D.: Building a mini-assembly system from a technology construction kit. Assembly Automation 24(1), 43–48 (2004)
7. Bi, Z.M., et al.: Reconfigurable manufacturing systems: the state of the art. International Journal of Production Research 46, 967–992 (2008)

8. Perremans, P.: Feature-based description of modular fixturing elements: the key to an expert system for the automatic design of the physical fixture. Advances in Engineering Software 25(1), 19–27 (1996)
9. Tsai, Y.-T., Wang, K.-S.: The development of modular-based design in considering technology complexity. European Journal of Operational Research 119, 692–703 (1999)
10. Levin, M.S.: Towards combinatorial analysis, adaptation, and planning of human-computer systems. Applied Intelligence 16(3), 235–247 (2002)
11. Hong, N.K., Hong, S.: Entity-based models for computer-aided design systems. Journal of Computing in Civil Engineering 12(1), 30–41 (1998)
12. Watson, I.: Case-based reasoning is a methodology not a technology. Knowledge-Based Systems 12, 303–308 (1999)
13. Tang, L., et al.: Concurrent Line-Balancing, Equipment Selection and Throughput Analysis for Multi-Part Optimal Line Design. The International Journal for Manufacturing Science & Production 6, 71–81 (2004)
14. Ohiro, T., et al.: A stochastic model for deciding an optimal production order and its corresponding configuration in a reconfigurable manufacturing system with multiple product groups. International Conference on Agile, Reconfigurable Manufacturing (2003); Ann Arbor
15. Zhao, X., Wang, K., Luo, Z.: A stochastic model of a reconfigurable manufacturing system Part I, a framework. International Journal of Production Research 38(10), 2273–2285 (2000)
16. Bi, Z.M., et al.: Development of reconfigurable machines. The International Journal of Advanced Manufacturing Technology (2007)
17. Shen, W., et al.: Applications of agent-based systems in intelligent manufacturing: An updated review. Advanced Engineering Informatics 20(4), 415–431 (2006)
18. Lohse, N.: Towards an Ontology Framework for the Integrated Design of Modular Assembly Systems. University of Nottingham (2006)
19. Zambonelli, F., Jennings, N.R., Wooldridge, M.: Multiagent systems as computational organisations: the Gaia methodology. In: Henderson-Sellers, B., Giorgini, P. (eds.), pp. 136–171. Idea Group Publishers (2005)

Reconfigurable Self-optimising Handling System

Rainer Müller[1], Martin Riedel[2], Matthias Vette[1], Burkhard Corves[2],
Martin Esser[1], and Mathias Hüsing[2]

[1] Laboratory for Machine Tools and Production Engineering (WZL),
Chair of Assembly Systems, RWTH Aachen University
[2] Department of Mechanism Theory and Dynamics of Machines (IGM),
RWTH Aachen University
{R.Mueller,M.Esser,M.Vette}@wzl.rwth-aachen.de
{Riedel,Corves,Huesing}@igm.rwth-aachen.de

Abstract. Demand for more versatile assembly and handling systems to facilitate customised production is gaining in importance. A new handling principle has been developed as a cost-effective approach to adapt to component-dependent tasks. It is based upon the gripping and movement of objects by multiple arms within a parallel kinematic structure. This structure combines the advantages of a system of co-operating robots with a simplified drive concept, in which the number of drives used is sharply reduced. On this basis, a modular assembly platform is being developed which, in addition to the kinematic units, also facilitates the integration of measurement, testing and joining modules. The modular concept also creates the conditions for a versatile, demand-driven layout of multiple kinematic units. This facilitates not only cooperative handling of large components using several gripping points, but also the transfer of objects handled between the individual units. These features of adaptivity are the basis for self-optimisation, which then can be implemented within a suitable control system.

1 Introduction

There has been a fundamental change in the conditions governing manufacturing industry in recent years. Progressive globalisation, rapid technological development and changes in the resources situation [1], [2] are responsible for increasing complexity and dynamics in industry and the industrial environment. One of the consequences is a further reduction of product life cycles, a sustained increase in the number of versions of products and constant pressure to cut manufacturing costs [3]. Assembly systems and processes are particularly exposed to these pressures, as they add a large part of the value in the manufacturing process.

2 Motivation

Progressive development in the field of industrial robotics has led to a variety of new applications in recent years, safeguarding jobs in high-wage European states by increasing the level of automation.

S. Ratchev (Ed.): IPAS 2010, IFIP AICT 315, pp. 255–262, 2010.
© IFIP International Federation for Information Processing 2010

Traditional demands on handling systems are currently undergoing change. In the past, higher and higher load capacity, greater and greater precision and higher and higher speeds have been demanded. However, priorities are increasingly shifting towards customised production and flexible solutions to component-dependent problems. Currently available handling systems frequently cannot fulfill the increased demands and complex tasks [4].

Against this background, one important current problem is the handling of large components, some of which have no intrinsic rigidity. Such components are used in aerospace systems, shipbuilding, wind turbine construction and the manufacture of solar panels. The trend is predominantly discernable in aircraft construction, where larger and larger shell elements in carbon-fibre reinforced plastics are being used for aircraft fuselages [5]. In the particular case of carbon-fibre reinforced plastics, care must be taken not to subject the components to any high forces during handling. Great precision and a large workspace are required. A single robot is not usually capable of moving the component without subjecting it to any forces. Large jigs are therefore used, rendering the system very inflexible and expensive.

Co-operating robots represent a more versatile approach to the problem described. This concept enables inflexible shapes to be dispensed with, as the component can be gripped and supported at different points by several robots [6]. The high procurement costs of industrial robots must be mentioned as a disadvantage. Complicated programming, which entails considerable set-up times, frequently means that such robot systems cannot be used cost-effectively.

Parallel kinematic structures are a cost-effective alternative. Even large, heavy components can be positioned quickly and accurately by closed kinematic chains. A disadvantage is the small workspace dictated by design, which is why parallel kinematic systems frequently do not represent an ideal solution to a general handling problem [7]. Therefore, a handling concept, which combines serial and parallel approaches, was designed.

3 Assembly Concept

An assembly platform which can be reconfigured for a specific task is required, particularly for large components in small series and with a lot of versions. The usual limits, such as the workspace and the range of components and use, have to be overcome to create an adaptable system which is reconfigurable and universally useable. A novel handling system has been developed. It is based upon gripping and moving objects by the use of multiple arms within a parallel kinematic structure. The parallel kinematic structure is regenerated with each gripping movement and integrates the object within the robot structure as a moveable platform, as shown in Fig. 1 [8], [9], [10], [11].

This approach combines the handling technology in known robot solutions to form a new overall solution, which is superior to the existing individual solutions in terms of reconfigurability, adaptability, cost-effectiveness and efficiency.

The main concept is based on the motion of a parallel robot, see Fig. 2. In this case, the object is bound to a moveable platform by a gripper, which is moved through the workspace by several kinematic chains. The joints on the platform combine the individual arms to form closed kinematic chains.

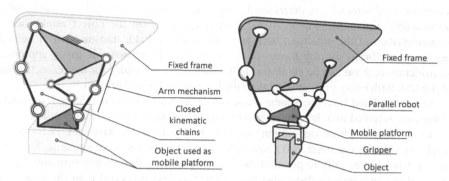

Fig. 1. New handling principle

Fig 2. Handling with a parallel robot

The arrangement and type of platform joints are essential to the kinematic structure, whilst the shape and size of the platform have no influence on the kinematics. If the joints are placed directly on the object handled, the object can be integrated into the structure as a platform and be moved by the system. The newly generated kinematic structure is similar to the structure of parallel robots. In contrast to cooperating robots, the configuration of the architecture in each object pose is determined by only six drives.

The handing method derived from this concept distinguishes three phases of movement, gripping, manipulating and depositing, shown in Fig. 3.

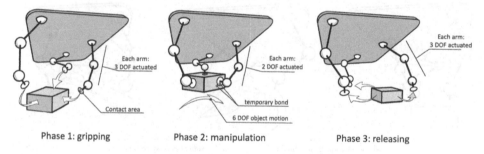

Phase 1: gripping   Phase 2: manipulation   Phase 3: releasing

Fig. 3. Movement sequence when handling an object

In the first phase (gripping), the object is free and the arms can move independently from each other to locate the contact elements on the object. These desired contact points are the later joint positions on the mobile platform. Three drives per arm are necessary to control the 3 degrees of freedom (DOF) of the individual arm. The contact elements are mounted swivelling and spring-centred in the wrist joints with additional 3DOF of rotation. At contact they align passively with the surfaces of the object.

In the second phase (manipulation), the contact elements create a bond with the object which can transmit all directions of forces and moments. The bond may be based upon various physical principles, e.g. form closure, temporary adhesion, sub-atmospheric pressure, magnetism, etc. The firm but pivoting bonds integrate the object into the structure and can now be moved as the latter's platform with 6 DOF. As the newly-formed structure is determined, only six drives are required for spatial object movement. Active positioning of the contact elements on the object establishes the characteristics of the structure, such as workspace, payload, transmission of velocity, accuracy, stiffness, etc, by means of the kinematic dimensions. If given appropriate consideration, these can be combined individually and optimally for the respective handling task with every grip, thus creating the basis for self-optimization.

In the third and final movement phase (releasing), the object is released, the contact forces are relieved and the arms are moved away, independently of each other.

In the second phase (manipulation), only two controlled drives are necessary for each arm, whilst the other two phases require an additional drive per arm. This additional drive serves only to position the contact elements during gripping and depositing and, as an auxiliary drive, can be designed as less powerful than the total of six principal drives which move the object. The secondary drives can be driven passively or decoupled in phase two. However, the drives can also provide active support to object movement if required.

Contemporary robotic solutions in which multiple integrated six-axis industrial robots move a component provide comparable payload capacity and versatility in use. Nevertheless, there are fundamental differences in drive configuration. The six drives in the new configuration contrast with the eighteen drives of a conventional "three-robot cell", entailing a considerable cost advantage for the new solution with comparable functionality.

The broad spectrum of use and adaptability of the new handling unit to changing tasks is achieved by two-stage convertibility of the system, Fig. 4.

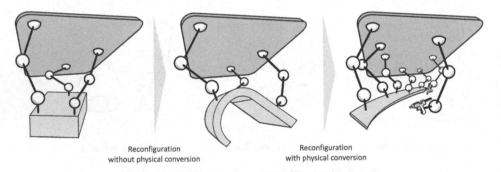

Reconfiguration
without physical conversion

Reconfiguration
with physical conversion

Fig. 4. Two-stage convertibility of the handling system

The first stage is adjustment of the kinematic dimensions of the platform to the imminent object movement. This adjustment can take place automatically on integration of the object into the structure by gripping. It does not require any physical conversion of the unit frame. This is a first step towards the goal of self-optimizing handling.

If the changes in the characteristics and dimensions of the objects handled are too great or if further functions such as measurement, testing or machining are to be integrated into the movement sequence, the system can be adjusted by conversion in a second stage. The base joints of the arms are mounted on a system of rails so that they can be moved easily. Additional kinematic units or modules can be integrated simply and quickly in a modular system.

4 Development Process

The development of such a handling system can traditionally be divided into analysis of requirements, concept, structural synthesis, dimensional synthesis, layout of secondary dimensions and design phases. The results of the concept phase have already been summarised in the previous paragraph. The subsequent developmental stages are described briefly below.

In an early stage of development, the kinematic structure is established so that the desired object movement is possible with 6 DOF.

Adapting Grübler's criteria demonstrates that the connectivity number of each arm must be equal to 6 to move objects freely within the workspace, regardless of the number of arms.

To identify suitable arm structures, it is expedient to break down the overall 6 DOF for each arm into a regional structure (3 DOF for translation) and local structure (3 DOF for rotation) and to look for partial solutions independently of each other. This procedure precludes structures with a purely parallel kinematic arm structure from the outset. These solutions would reduce the workspace of the arm assembly significantly.

A systematic approach facilitates full inclusion of the solutions for the partial structures with 3 DOF in translation. Both serial and parallel kinematic chains as well as combinations of both (hybrid structures) are considered. The comprehensive host of solutions is reduced to a small number of possible structures by restrictions in a preselection process. They are analysed individually and assessed using the criteria in the process compliant with VDI 2225.

The finally selected kinematic structure is shown in Fig. 5.

A detailed description of the development process and the kinematic computation of the structure are given in [8], [9], [11] and [12].

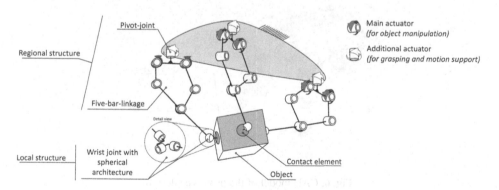

Fig. 5. Selected kinematic structure

After all the principal parameters were determined, the other components were projected and designed in detail on a CAD system. Questions such as component layout, types of connection and suitability for fabrication and installation are considered, producing the prototype design shown in Fig. 6.

The prototype consists of three arms pivot-mounted on the fixed frame so that they can rotate vertically and articulated to the object during the latter's movement. The fixed frame measures approximately 1300 x 1500 mm² (W x D) and allows the arms to be applied in different configurations.

Each arm is 900 mm long in its extended state and consists of a five-bar linkage parallelogram as the guidance mechanism with a tilting frame. This tilting frame carries both the servomotors which drive both cranks of the parallelogram by belts. The drives are located so that their inertia effect is minimal in an acceleration of rotation of the arm about the vertical axis. The vertical axis is also driven by a servomotor/belt combination, the size of which can be reduced by the mitigated demands of the gripping function. The five-bar linkage moves the wrist joint with the contact element. In initial prototypes, an electromagnet with a weight of 300 g is used, capable or generating a temporary bonding force of 700 N. A gas-pressurised spring relieves the drives when lifting heavy components, thus increasing the load-bearing capacity.

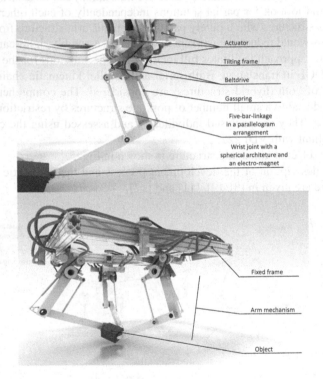

Fig. 6. CAD model of the prototype design

5 Application

The objective is the production of a platform concept which can be configured to the task. Beside the kinematic units, further modules are to be integrated. They include measurement and testing modules, and modules for joining and assembling components.

Additional units and drives can be used to statically over-determinate the system and keep the work piece free of kinematic load.

One specimen application of such an assembly platform is stringer installation in aircraft construction. In this case challenges arise predominantly from the spherical curvature of the shell components and the accompanying stringers of different sizes and curvature which have to be handled. A further difficulty is that the stringers, which are up to 15 m long, are limp and therefore have to be supported by additional grippers. Modularisation enables the kinematic units to be positioned suitably.

Cooperative handling produces a further advantage. A stringer can be passed on to another kinematic unit before it leaves the workspace of a kinematic unit. Similar concepts are also used in cooperating robots, in which parts are passed from one robot to another. Large workspace can then be covered by relatively small handling units. Such transfer of components cannot usually be achieved cost-effectively by conventional industrial robots, due to the high number of kinematic units required.

6 Self-optimisation

The degrees of freedom of the introduced system offer adaptivity for a wide spectrum of handling tasks. Nevertheless, they also imply a large planning effort for finding and optimizing the configuration needed for an individual handling task. Thus the benefit of adaptivity is reduced. Self-Optimizing, i.e. automatically generated configuration for the best possible handling solution is necessary for the cost-effective use of the assembly platform's reconfigurability.

The information needed to perform automatic optimization are the handling task (shape and path of the part) and the configuration of the handling system.

The user will have the opportunity to specify the path in a virtual 3D-model and simulate the process after automatic calculation of the optimal configuration. Reconfiguration of the moving platform is then performed automatically by re-gripping the part at the proposed locations. Reconfiguration of the number and base points of kinematic units will be done manually. After this, the precise kinematic parameters for the assembly platform can be identified automatically using an external measuring system, e.g. iGPS. This has the advantage of reducing set-up time further as the kinematic units do not have to be aligned precisely by hand.

The handling motions are then calculated by using the real geometric parameters of the system and the given path of the work piece. Additional manual adjustment of the path, as in conventional off-line programming, can largely be waived.

7 Summary and Outlook

A reconfigurable handling concept has been introduced that combines the advantages of cooperating serial robots and parallel kinematics to overcome the use of large fixtures when handling small series of large parts. The design and application of the concept has been shown and it was made clear, that the reconfigurability can offer significant economical advantages, when the increased planning effort is addressed by means of self-optimization. Based on the mechanical and control development described within this paper, the chosen concept of self-optimization will be further developed and implemented.

Interface management constitutes a further important aspect. To facilitate development of an adaptable assembly platform, it must be ensured that other objects can also be integrated, as well as the developed kinematic units. Further handling, control, measurement and test systems are conceivable. For this reason a modular open control system has been selected for future expansion of the system.

The developed kinematic units, existing conventional robotic systems, measurement systems, etc could be integrated to meet demand practically without restrictions and used in a single system against a background of consistent mechatronic modularisation, on the basis of the common platform.

The depicted research has been funded by the German Research Foundation DFG as part of the Cluster of Excellence "Integrative Production Technology for High-Wage Countries".

References

1. Möller, N.: Bestimmung der Wirtschaftlichkeit wandlungsfähiger Produktionssysteme. Forschungsbericht IWB, Band 212 (2008)
2. Müller, R., Buchner, T., Fayzullin, K., Gottschalk, S., Herfs, W., Hilchner, R., Pyschny, N.: Montagetechnik und –organisation. Apprimus Verlag, Aachen (2009)
3. Lotter, B.: Montage in der industriellen Produktion. Ein Handbuch für die Praxis. Springer, Berlin (2006)
4. Nyhuis, P.: Wandlungsfähige Produktionssysteme. Heute die Industrie von morgen gestalten Garbsen, PZH, Produktionstechn. Zentrum (2008)
5. Licha, A.: Flexible Montageautomatisierung zur Komplettmontage flächenhafter Produktstrukturen durch kooperierende Industrieroboter. Meisenbach, Bamberg (2003)
6. Feldmann, K., Ziegler, C., Michl, M.: Bewegungssteuerung für kooperierende Industrieroboter in der Montageautomatisierung. wt werkstattstechnik online. Jg. 97(9), S.713 (2007)
7. Rückel, V., Feldmann, K.: Komplettmontage mit kooperierenden Robotern. wt werkstattstechnik online Jg. 95(3), S. 85 (2005)
8. Riedel, M.: Systematische Entwicklung und Konstruktion einer Handhabungseinheit zum Greifen und Manipulieren von Objekten auf Basis einer parallelkinematischen Struktur. Institut für Getriebetechnik und Maschinendynamik an der RWTH Aachen (2005)
9. Riedel, M., Nefzi, M., Corves, B.: Development and Design of a multi-fingered Gripper for Dexterous Manipulation. In: Mechatronic 2006, 4th IFAC-Symposium on Mechatronic Systems. in Eng., Heidelberg, September 12-14 (2006)
10. DE 10 2005 059 349.6, Parallelkinematisches Greifsystem zum Greifen und Handhaben, PARAGRIP (2005)
11. Riedel, M., Nefzi, M., Hüsing, M., Corves, B.: An Adjustable Gripper as a Reconfigurable Robot with a Parallel Structure. In: Second International Workshop on Fundamental Issues and Future Research Directions for Parallel Mechanisms and Manipulators, Montpellier, France, September 21–22 (2008)
12. Müller, R., Corves, B., Hüsing, M., Esser, M., Riedel, M., Vette, M.: Rekonfigurierbares selbstoptimierendes Bauteilhandling. In: 8. Kolloquium Getriebetechnik Aachen 2009, Verlagshaus Mainz, Aachen, pp. S.297–S.311 (2009) ISBN: 3-86130-984-X

Modular and Generic Control Software System for Scalable Automation

Christian Brecher, Martin Freundt, and Daniel Schöllhorn

Fraunhofer Institute for Production Technology, Steinbachstrasse 17, 52074 Aachen, Germany,
Tel.: +49(0)241-8904-253, Fax.: +49(0)241-8904-6253,
martin.freundt@ipt.fraunhofer.de

Abstract. The development of automated production systems is subdivided in two mayor tasks. One is the development of the processes needed to meet the requirements for the product, the other is the setup of a control system enabling the hardware to perform these processes. Typically the larger amount of the available resources is needed for the setup of hardware and implementation of the required control mechanism, leaving only limited resources for the process development. Especially for small scale and prototype production with a high rate of changes, this is why fully automated solutions don't pay off and manual or partial manually assembly is preferred [1]. This paper introduces an approach how to separate the implementation effort of the hardware specific tasks from the process definition, allowing a fast and easy setup for new automation systems, due to simultaneous development and a high rate of reuse of previous solutions.

Keywords: Software, framework, control system, one piece flow, assembly.

1 Introduction

For most R&D orientated Enterprises as well as high tech production sites the implementation of automated processes is dominated by fast changing equipment and complex production or test setups. To be able to cover various automation tasks, a wide range of equipment (e.g. Sensors, Robots, Actuators) has to be integrated within the assembly system [2]. In order to make different types of hardware work together, engineering the control software usually takes most of the time. Even when components have been used in automated setups before, the interfaces have to be adapted to each other and the particular process requirements. As a result, a lot of work is required to integrate the same hardware again and again just because the setup of the hardware configuration has changed.

Even the control programs for the realization of the processes are developed specific to the particular application without the option to efficiently prepare for a future reuse of software code due to the interdependences of process and hardware. According to the experience with hardware setup and process development the effort needed for automation was about 70 % of the overall time leaving only 30% of the resources for the actual task, the process engineering.

To be able to focus on the process engineering it is necessary to decouple hardware setup and process development, at least in regard to the software programming. A

S. Ratchev (Ed.): IPAS 2010, IFIP AICT 315, pp. 263–270, 2010.

new approach within software for the automation of processes is required. This software system must enable a full reuse of previously developed solutions of hardware specific code as well as code describing the process flow. Once developed and implemented, a process like gripping a part in a special way or moving a robot towards a position and orientation, should be a procedure that does not require additional programming when used in another context. The same applies for the code providing hardware specific functions, like taking care about signal generation, positioning processes and communication with controllers.

The aim is to program the hardware typical functions within "hardware specific" and "process specific" code separated from each other. Thereby the point of view has to be changed form the conventional thinking "what the robot has to do in order to make the process happening" towards focusing on "what has to happen with the parts in order to assemble them". Thereby, it doesn't matter which kind of hardware configuration will be used to implement the process description. In addition, the software has to be extendable to allow a fast and easy integration of new components.

Based on this vision, a software system was designed and realized, achieving the separation of process related and hardware related code elements. Therefore the control system has to be grouped in different divisions representing characteristic aspects of the overall system:

- **Hardware representation:** Encapsulating all properties and functions of the available hardware, including data communication and control of hardware internal functions
- **Process representation:** All process specific elements have to be encapsulated within a code division, so that reuse is independent form the hardware components which are utilized for execution.
- **Basic system functionality:** The interaction between the hardware specific code elements and the process specific code has to be organized by a component. This component has to ensure that the process description will be put into action as adequate hardware is available.

Main condition is, that these software divisions are independent from each other although they need to interact in order to perform the designated tasks. The key benefit of this approach is the complete separation of the hardware control programming effort from the process description programming effort. In doing so, it is possible to setup and configure the processes without the need of knowledge about the hardware of the system that performs the later actions. Furthermore, the hardware control can be configured without the knowledge of the processes allowing a separated development process.

This is a key feature for a fast and efficient process development as existing solutions like gripping a component or aligning a component based on a reference signal can be reused without code implementation.

2   Application Scenario AutoMiPro

To realize the development and implementation of a software system with the abilities and specification described above an adequate application scenario is required. Inside the AutoMiPro research project at the Fraunhofer IPT, logistics in terms of the design of hard- and software solutions for a changeable production respectively a one piece material flow within automated precision and micro assembly processes was developed. A demonstrator was rigged at the Fraunhofer IPT, consisting of four main components shown in figure 1. The skills of component processing are demonstrated within a diode laser assembly process containing various micro optical components. Therefore, hardware is needed capable of handling parts with different shapes, materials and handling requirements. As micro mechanical and optical parts of diode laser systems differ in their parameters due to the production process, flexibility is required. In order to increase efficiency by minimizing the use of sensor systems, every part must be treated as a unique entity with its corresponding data sheet characterizing the part. Thus it is possible to allow for handling and positioning of each part based on its real properties within a distanced process. In addition, information generated during the production process can directly be stored part specific, enabling a full traceability of each production step. The assembly process of a diode laser system itself is very complex. Alignment of at least one, often more degrees of freedom is required for most of the optical parts. The alignment starts to get more complex as different alignment movements within the multiple degrees of freedom affect each other. By using the unique data for every part it gets possible to apply adjustment processes on different parts or at least part variations without change effort, once part properties are used to parameterize the alignment processes.

In order to allow for a company-overreaching production it gets even more important to separate the process description for hardware and form part specific properties and functions. As the production chain overreaches different hardware equipment and technology, a hardware independent description of the processing is required. This hardware independent process information is linked to parts itself, the same way the properties of the part are stored. The part so called "carries its own assembly description" through the production process. Assuming that all hardware control systems are capable of using this information, the necessary activation of hardware can be generated

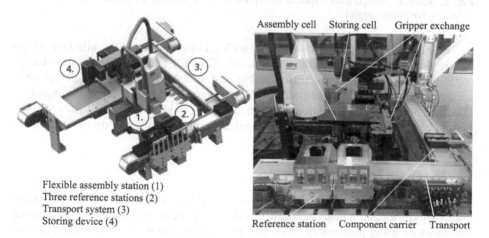

Assembly cell   Storing cell   Gripper exchange

Flexible assembly station (1)
Three reference stations (2)
Transport system (3)
Storing device (4)

Reference station   Component carrier   Transport

Fig. 1. Demonstrator setup of the AutoMiPro Project – Application: diode laser assembly

automatically inside the software. In this case no additional input to the system is needed, realizing a clear separation between the hardware control issues and the product, respectively the process leading to the product.

3 Design of the Software System

To implement a software system that completely separates the programming of the hardware control form the specification of the process information two different types of software fields are needed. One for the hardware and its capabilities and one for the process information describing parts and procedures. The framework organisation element has to be capable to separate and organize the interaction between the hardware and the process description. In addition it must allow for flexible, fast and easy reconfiguration of the current hardware setup. Those three components, represent all aspects of a fully functional assembly system. (ref. figure 1).

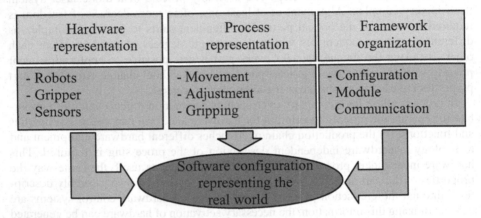

Fig. 2. Software components required to represent all aspects of an automated process flow including hardware control

To equip an organisation component with an appropriate logic, allowing to organize module interaction, causes additional effort for configuration. It also handicaps the goal of being universally usable and scalable as configuration specific, application related code is required [3]. Due to a required reduction of complexity it is essential, that the organisation component do not have to be aware of the specific system constellation in order to avoid the described effort for an omniscient organisation component.

3.1 Hardware Representing Modules

The main goal regarding efficiency within the setup of an automated device is reusability of known and already realized functions, like accessing hardware via external interfaces or using positioning strategies for a robot system. Further, being able to apply changes within a running system without retesting and modifying further system programming is important to extent and optimizes functionalities.

Conventional control systems, often decentralized on robot controllers are highly integrated due to the limited abilities of the programming language or fundamental execution of NC-Code. All hardware components therefore have to be controlled by one device allowing for accessing all data and being able to integrate flexibility. In order to be able to use knowledge and skills of e.g. computer science personal, a PC platform and a common high level language like C#, capable to realize object orientated software has to be utilized.

The object orientated programming allows for separation of properties and functionalities into separate software objects, each representing the smallest entity of a hardware component. A device with an integrated pneumatic actuator for example would be separated into the device specific part and the pneumatic actuator. This actuator itself is connected to a standard component, a pneumatic valve, which is already available and completely represented within the software framework. Thereby the functions for switching valves and accounting for actor properties, already is included within the software and can be reused.

In order to minimize the effort of extending the system with additional objects the method of inheritance of software objects is used. Inside the developed software structure every part of the real world is represented by a single software object, containing all information and providing all functions of its real-world counterpart.

By strictly separating and isolating functions and properties into software objects, it gets possible to modify and even replace whole elements with no or at least minimal effort. Communication between these objects is realized by primitive commands respectively messages. Therefore each object is programmed to be capable to process certain messages and commands.

In order to avoid the implementation and configuration of an omniscient organisation component, the software objects, representing hardware components, are organized within a hierarchical, tree like structure. This structure between the software objects has to be defined at system start up based on the relations of the real system configuration. The system automatically keeps it up to date during automated processing. With this approach production hardware as well as product parts can be described the same way. Figure 3 illustrates how a gripper carries a part, in this case a Fast-Axis-Collimation lens. The gripper is subordinated underneath the bottom side of the gripper interface, which is connected to the upper interface side. The upper interface is linked to the robot taking care about its positioning.

All information concerning the position and orientation of components is defined relative to the superior component. This allows for a fully automated update of the data, allowing automated configuration modifications like gripper exchange or the gripping and releasing of parts. Therefore the part is always dedicated to the component defining its position within the real setup. If it is placed within a fixture, its position is defined relative to the fixture. In case a robot respectively a gripper handles it, it is dedicated relative to the gripper, which defines its current position. In order to extract the information about position and orientation of a component, the part can be "asked" for its position. Once this request is started, internal functions all objects provide, causes the calculation of the absolute position based on all relative data independent to the current configuration of the system.

In order to ensure proper precision of the virtual model, certain process steps can be used to update the relative position of a component. A fully traceable production can be realized linking all historical data and part information within the data set for the assembled product. Even relevant process data can be collected within processing and assigned to the processed components for full traceability within conventional production as well as for process ramp up or process development.

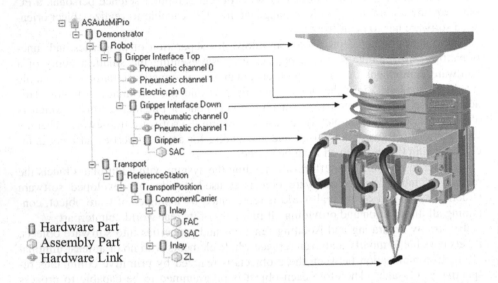

Fig. 3. Example for the representation of real world elements within the software system

In addition, the modular concept of the software control guarantees an easy exchange or modification of hardware components or process logics. If for example a new gripping device is added, only the gripper software object needs to be exchanged. Therefore, an existing gripper class definition can be derived to a new gripper subclass or only be parameterized without adding further functions. All other participating modules, controlling the other hardware and processes, stay as they are, with the ability to use the new gripper right away. Thus a toolbox of different software modules is created. The operator is allowed to simply select and connect the required modules and configure the hardware setup as needed.

3.2   Process Representing Control Modules

The second software element addresses the way process information is described. Due to the modular and flexible setup of the assembly, the description of the processes has to be hardware independent. This is crucial for parallel development of hardware and process as well it is a basic condition for reusability of developed process steps. Therefore, all process descriptions are defined parameterized. Parameters can be software objects representing hardware components like grippers or representations of objects being processed as well as simple parameters. All absolute values, in particular position and orientation values, required for the execution of handling processes are thereby defined based on the component data, calculated on the fly once the process description is executed. This ensures that current configurations and positions of robots or other variegating properties are included once a command or process description is executed.

On behalf of a deterministic process development, detailed process descriptions can be used for defining the processing based on elementary commands, like "move left 2 mm", with no margin for variances. Due to changeability demands and the need for simplification of the process development, more abstract commands like "position part A next to part B with an offset of 2 mm" can be executed and realized. For more complex logics, as alignment processes, a process command object is available allowing for encapsulation of process logic by the generation of command primitives, executable by the software framework.

4 Process Example – Diode Laser Assembly Process

To adjust a micro optical lens in front of the diode laser source, multiple different procedures have to be executed. As described, the information specifying these procedures are stored as a list of components dedicated to the part information of a so called master part, describing only the "what is to do" but not the "how to do it". This includes command primitives and complex process descriptions like adjustment and alignment functions e.g. for the fully automated active alignment of a Fast-Axis-Collimation lens. As the assembly process is started, for example the FAC lens is asked to move itself towards the diode laser in, in order to get it into adjustment position as shown in figure 5. For this operation it is completely regardless where the part is located or how it will be transported. The part, supposed to move itself just creates a new command to provoke its transport to the target position, as it is not capable to move itself. This command is always addressed to the superior software object until a software object is identified, accepting the command for execution (ref. Figure 3). In case of the linking of a positioning task, each rejection of the command causes a modification in order to accomplish the initial positioning request. As the part wants to be moved and is not capable to move, it asks the gripper, currently defining its position to move. The gripper, itself not capable to move, recalculates the moving request in a way that it wishes to get moved by its superior object so that the FAC lens will be moved the way wanted to, initially. By this, a command is transferred through the hierarchical system, although no component is aware of the overall structure, until a software object accepts execution of the current command. By this means, boundary conditions like acceptable gripping methods or gripping force limitations can be accounted for in an unknown environment of hardware components. This enabled for parallelisation of development and extendibility of already producing systems but also cause only partial predictable process realisation.

Although it is possible to assign commands to be executed by predefined hardware components, using their unique object ID, instead of using hardware independent commands. This option keeps the concept scalable for development and introduction within industrial applications; even some of the advantages of the concept thereby get reduced. Similar to parts provoking their own processing by creating command primitives, process objects itself can create requests to get for example information out of a vision system as it is required for the handling and alignment of a FAC lens as shown in figure 4.

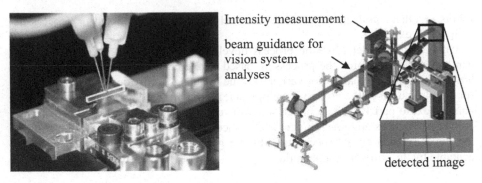

Fig. 4. Fully automated FAC alignment at the Fraunhofer IPT using hardware independent process descriptions and process independent hardware programming

5 Summary and Outlook

The presented software framework is currently in implementation. Defining hardware-independent positioning processes and executing theses on different hardware configurations is already put into application. Further expansion of the system by integration of additional devices is ongoing. As the designed system provides a solid base for complex automation tasks, additional research will be conducted within the field of the design of more complex process objects in order to allow for automated rigging and adaptive or self learning processes.

Acknowledgements

The authors thank the German Federal Ministry of Education and Research (BMBF) within the project »AutoMiPro«, part of the Framework Concept »Research for Tomorrow's Production« (02PB3140) managed by the Project Management Agency Forschungszentrum Karlsruhe, Production and Manufacturing Technologies Division (PTKA-PFT).

References

1. Hesselbach, J., et al.: mikroPRO – Untersuchung zum internationalen Stand der Mikroproduktionstechnik. Vulkan-Verlag, Essen (2002)
2. Robot Visions to 2020 and beyond – The Strategic Research Agenda for robotics in Europe (2009)
3. Bara, J., Cararinha-Maos, L.: Coalitions of manufacturing components for shop floor agility. International Journal of Networking and Virtual Organisations, IJNVO (2003) ISBN 1470-9503

Agile Multi-parallel Micro Manufacturing Using a Grid of Equiplets

Erik Puik, Leo van Moergestel

Utrecht University of Applied Science, Faculty Nature & Technology,
Micro Systems Technology, Oudenoord 700,
3513 EX, Utrecht, The Netherlands

Abstract. Unlike manufacturing technology for semiconductors and printed circuit boards, the market for traditional micro assembly lacks a clear public roadmap. More agile manufacturing strategies are needed in an environment in which dealing with change becomes a rule instead of an exception.

In this paper, an attempt is made to bring production with universal micro assembly cells to the next level. This is realised by placing a larger number of cells, called Equiplets, in a "Grid". Equiplets are compact and low-cost manufacturing platforms that can be reconfigured to a broad number of applications.

Benchmarking Equiplet production has shown reduced time to market and a smooth transition from R&D to Manufacturing. When higher production volumes are needed, more systems can be placed in parallel to meet the manufacturing demand. Costs of product design changes in the later stage of industrialisation have been reduced due to the modular production in grids, which allows the final design freeze to be postponed as late as possible.

The need for invested capital is also pushed backwards accordingly.

Keywords: Grid Manufacturing, Agile, Reconfigurable, Equiplet, Production, Micro, Microsystem, Hybrid, Submissive Product Design.

1 Introduction

This research focuses on manufacturing of "Hybrid Microsystems". Hybrid systems are composed of multiple elements to realise their primary function [1, 2]. Hybrid Microsystems combine multiple functional domains; electronic, mechanical, optical, fluidic or biomedical domains may be included. Though all these different systems share many of the same underlying technologies, the use of micro technology is typically application specific and mainly used for sensors and actuators. This is the reason why Hybrid Microsystems address, a limited number of mass applications excepted, mainly niche markets.

Manufacturing of products for niche markets with numbers in the mid size volume area, in between 10^4-10^7 products annually, is not easy from an efficiency point of view. Dedicated equipment for manufacturing requires high added value at these relatively small production series. Manual production on the other hand strongly relies on labour, which not only is expensive, but also could suffer from quality issues in the high-tech systems markets with its strict quality demands.

S. Ratchev (Ed.): IPAS 2010, IFIP AICT 315, pp. 271–282, 2010.

2 Manufacturing Context for Micro Systems

2.1 Disruptions During Ramp Up of Production in the Micro Domain

Manufacturing accuracies for assembly operations in the micro domain are constantly evolving. Since micro system manufacturing, in the lower and mid volume domain, is traditionally relying on an extensive amount of manual labour. Supporting assembly tools are typically used, to enable manual workers, in meeting the high accuracies of production. Once fully ramped up, production equipment has been developed and produced. In the move from early production to the ramped up situation, a transition is made from manual production to (semi) automated production. This transition, with large impact on process stability and product quality, is made at some point in which market introduction is at sight, a moment when *invested capital and project risks are rapidly increasing*. Unexpected delays at this point, due to the disruptions of this manufacturing transition, will almost certainly lead to *delayed market introduction* [3].

To avoid the disruptions during ramp-up, applied equipment is preferably to remain unchanged during the transition from R&D to full production. This would be the case if assembly tools during R&D stage, or at least the assembly principle and the conditions under which they are performed, could be reused for semi automated production and later eventually even automated production. Attempts have been made in the past to realise this way of working [4-7].

2.2 Re-configurability

Unlike manufacturing technology for semiconductors and printed circuit boards (PCB), the market for traditional micro assembly lacks a clear public roadmap [8-10]. It is due to the larger variety of processes and the three dimensional characteristics of micro assembly that the technology will not fit into a single manufacturing framework. Due to the high added value of many manufacturing processes, dedicated manufacturing systems have proven to be legitimate in micro manufacturing, even if the manufacturing technology is intrinsically less efficient. On the other hand, many products with lower economic potential don't make it to the market because for this reason. It is the potential of reusing existing manufacturing technology that increases the popularity of flexible equipment. Many suppliers of state of the art equipment offer systems that can be configured in a flexible way to form a dedicated manufacturing system. Problems however occur when these systems have to be converted to changing products. Usually machines are *shipped back* to the factory to be modified (upgraded). This *ceases actual production*, which generally is not acceptable. Therefore the choice is often made to replace the equipment and invest in new systems, leaving the existing systems untouched till the new system is up and running. *If existing equipment would be gradually upgradable, in a true re-configurable sense, investments in equipment could be reduced significantly.*

2.3 Agile Manufacturing

The manufacturing environments of present and next decades differ substantially from the past. Technology and consumer markets have become extremely difficult, posing difficulties for the use of traditional top-down planning-based methods for developing manufacturing strategies. More agile manufacturing strategies are needed in an environment in which *dealing with change* becomes *a rule instead of an exception* [11].

Manufacturing Technology is regarded as one of the most important decision areas within the manufacturing management function. Traditionally, management has influenced manufacturing technology to a much greater extent than the other way round. To have a revolutionary manufacturing concept adopted, one of the aims of this manufacturing strategy is to give the organisation *strategic direction with regard to manufacturing issues*. This direction should include technology, people and infrastructure, used in a consistent way, with the strategic objectives of the business [12, 13]. In a future world, where change is a certainty, all company's resources should work together closely to deal with unplanned change. Basically there are only two ways of handling this, *being in control or being flexible*. In this context it is hard to imagine how traditional dedicated equipment would suffice the needs for the future.

2.4 Submissive Product Design

To improve reuse of technology, during the transformation from R&D to full-scale manufacturing, a distinct conformity has to be realised between product design and manufacturing processes [14]. This could lead to more cost effective, gradual and low-threshold industrialisation of micro systems [7, 8]. Basically, this way of working is based on the application of design rules that have been strongly submitted to the available production framework of the manufacturing facility. *Submissive Product Design can be seen as a fortified implementation of DFA* (Design For Assembly) in the micro domain. With Submissive Product Design, *manufacturing technology is unceasingly leading over the Product Design*. If applied conformably through the industrialisation processes, manufacturing complexity will be disentangled into basic assembly operations [8]. This means the use of a larger number of machines with less complex assembly operations. The assembly operations would still require leading edge manufacturing technology, but less assembly research is needed to implement the technology in manufacturing equipment. This paves the opportunities for standardisation, and the much aimed for, reuse of manufacturing technology.

3 Multi Parallel Micro Manufacturing; "Grid Manufacturing"

3.1 Production in the Micro Domain Using "Equiplets"

Typical required accuracies in the micro assembly domain exceed the achievable accuracies of manual labour when performed without any some of instrumentation. With straightforward assembly tools a significant gain in quality of high performance assembly operations can be achieved. The application of manual labour has disadvantages too, mainly in achieving constant & guaranteed accuracies. This preferably would limit manual intervention of operators to the logistical part of the assembly, like handling parts in trays [15, 16]. *The need for increased production conformity, or yield, leads to the implementation of (universal) micro assembly cells.* Much effort has been performed to design these universal assembly cells that meet the agility of human operators [15, 17-27]. Secondly, cost analyses have been performed to justify the application of these cells in an economical context [16, 28, 29]. These efforts have been successful, in a way that universal micro assembly cells can be feasible from a technological and economical perspective. The throughput of these systems however has not exceeded mid-volume production quantities. *In this paper, an attempt is made to bring production with universal micro assembly cells to the next level.* This is done by placing a larger number of cells in a "Grid" (4-100$^+$ systems). The universal assembly platforms, which are field reconfigurable, are called "Equiplets". Equiplets are compact and low-cost manufacturing platforms that can be reconfigured to a broad number of applications by adding product dependent tools. These tools are very similar as the tools that are used during R&D and for the manufacturing of pre-production series.

Using Equiplets in production has advantages and drawbacks. These are compared in the next chapter. Two advantages may be seen as the main drivers for applying Equiplets:

- Reduction of risks during Ramp-up. The transition from "R&D-Pilot Production-Automated Production" can be made without disrupting production technology. Optimal consistency through these phases is achieved, minimising risks of non functioning production technology;
- Flexibility gain in the application of production resources. Due to the truly reconfigurable character of Equiplets, and also due to the larger number of Equiplets present in a factory, it is possible to "balance" regular production with the actual production demand. This could be used to balance different production stages of one product, but can also be used to balance production demand over a number of different products to be produced at the same factory.

Picture 1-3. Equiplets in R&D environment. Equiplets are easy to reconfigure due to the modular structure. Ready for production (left) or for a quick test (right), the structure allows experimenting. With an agent based control system, self-learning is an option for the future to further increase accessibility (middle)

3.2 Dynamics of Manufacturing in Grids

Analogue to "Grid Computing", where time consuming computer tasks are divided over many computing hardware systems, it is also possible for manufacturing tasks to be distributed over a larger number of production systems. It may be clear that, due to the character of micro assembly, the dynamics of tuning the grid to adapt to changing demand for manufacturing will be of another order than software computing grids. Grid refinement for manufacturing will more likely be on a daily basis (compared to seconds in grid computing). This however is a breakthrough in flexibility compared to dedicated equipment, where substantial equipment upgrades are in the magnitude of months.

Manufacturing tasks are very different from computing tasks in a sense that the adapting the manufacturing grid to start another product cannot always be done in software alone. Note that re-configurability in this context does not mean that the production system is suitable for all thinkable production demands. This is the main reason why manufacturing in grids can only be successful if combined with a solid design strategy like DFA or better "Submissive Product Design". Although some product switches, especially switches within the same product family, can be done by changing the software jobs alone, it is however most likely to have a combined software and hardware modifications.

Modern adaptive controller architectures, based on software agents, support flexibility and speed of re-configuration, this will be addressed in a future paper.

4 Benchmarking Equiplet Production with State of the Art Manufacturing Equipment

In this chapter, Grid Manufacturing with Equiplets is compared with state of the art manufacturing solutions on the following topics:

- Time to market;
- Investment risks;
- Production agility;
- A last topic, to be covered is "Cost Modelling" research. Some basics will be explained in this document, a deeper cost modelling analysis is planned for a near future publication. This will cover issues like cost of ownership, but also up front investing in flexibility that might payback later.

4.1 Time to Market, Time to Volume

Industrialisation of new product developments usually starts with a limited number of R&D prototypes, realised by R&D engineers, to mainly serve for testing of functionality of the product. R&D engineers are usually no experts on manufacturing, which causes the need to optimise the product design for industrialisation. The traditional approach, till some 20 years ago, tended to be fully sequential, but since the 90's, concurrent engineering strategies have appeared to be more efficient. The overlap of R&D "proof of concept" and "manufacturing engineering" leads to quicker industrialisation, bringing problems due to the up scaling process to the surface in an earlier stage. This while manufacturing investments are still low thus keeping financial risks acceptable. The early involvement of manufacturing engineers leads to more optimised manufacturing processes that are upgradable to reasonable extent. Traditional high volume production with dedicated equipment however, would still need a redefinition of production process and equipment, causing the need for an "equipment feasibility" phase followed by a certain timeframe in which equipment is being engineered and realised (figure 1). Pilot production cannot start before these phases have been completed.

If however the manufacturing of prototypes could be structurally upgraded by the introduction of robotised equipment, manually- or tele-operated in the R&D phase, the technological basis of the production processes would automatically meet the demand for automated manufacturing. In the transition to pilot production, human commands would be gradually replaced by software automation, offering a seamless path to pilot production. As a rule of thumb, it should be envisioned from a manufacturing point of view, that it is not desirable to perform manual operations to prototypes during the R&D phase, but that the operations always should be performed by equipment, or simple reconfigurable equipment, which leads to the use of Equiplets during the entire development and industrialisation phases. Without this dedication the advantages will be gradually reduced to a point in which industrialisation is typically performed sequential, analogue to today's standards.

Time Path to Market

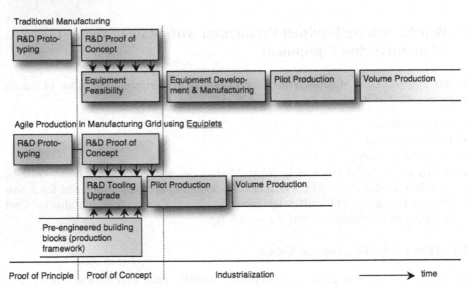

Fig. 1. Concurrent development and engineering of High Volume manufacturing processes during the late proof of concept phase. By upgrading well-chosen production technology from R&D to production, a significant time gain can be achieved in bringing new developments to the market.

It will need no explanation that manufacturing in a grid works best when the attitude of R&D engineers is cooperative. As stated in chapter 2, this calls for an increased level of management involvement, which should aim on technology and people as well as the necessary infrastructure. A positive stimulant, to optimise R&D involvement with Equiplets during early phases, is the energy it usually takes to realise prototypes and demonstrators in the early development stage. Especially in the micro domain, this usually absorbs significant amounts of resources. By using robotised equipment in the early stage, the numbers produced could be increased at minimal cost, providing more and also better quality prototypes supporting the research process.

The last point of attention is the process of ramping up to higher volumes. In a traditional way this would be typically done by increasing uptime and the speed of the dedicated production machine. In the case of Equiplet production this is basically the same, under the constraint that the maximum speed of a Equiplet production line will be structurally lower than the well engineered and state-of the art built production machine. This would mean that when the maximum throughput of one production line has been reached, the next step would be to increase the number of production systems that operate in parallel. A major advantage is that the investments in more parallel production lines only has to be sustained when the market demand indeed reaches the next level. If a product appears less successful or has been delayed, according investments can be postponed. If a product is extremely successful, more machines can be configured on fairly short notice, since the complexity of the Equiplets is low due to the decomposition of technology over a larger number of machines. In this way, a manufacturing plant always fits the actual demand (figure 2).

Gradual Expansion to meet Volume Demand

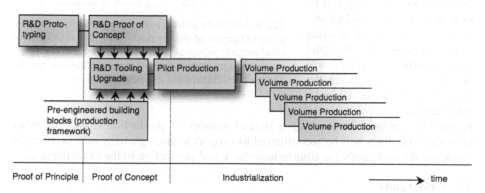

Fig. 2. Volume production can be adjusted to market demand, requesting investments tot be done no earlier than revenues start increasing. In this way, investment risks are minimised.

4.2 Investment Risks

In chapter 2.3 was stated that manufacturing technology is regarded as one of the most important decision areas within the manufacturing management function [11]. Though this true, and there is no shortage on management involvement for manufacturing departments, the moment that the involvement takes place is often timed badly. Management involvement typically increases when the investments are on the urge of rapid increase. This is when R&D is in the Proof of Concept Phase [30] (figure 3). The problem however is that at this point many design issues, and related manufacturing demands, have already been chosen and the costs for industrialisation have been intrinsically determined.

A basic difficulty with manufacturing equipment is that the manufacturing means have to be determined developed and invested up front. With the engineering and realisation of manufacturing equipment, the product design will be frozen except for the exact specified (foreseen) variations within a product family. This is the main reason that the "Cost of Change" for dedicated equipment, as given in figure 3, increases so quickly. Once production has started, problems even get worse, because the manufacturing process has to be shut down to bring product modifications in effect, putting pressure on revenues.

How different will manufacturing be if possible to incorporate a growth strategy with parallel production platforms. With manufacturing in grids this can be achieved after product launch to grow to a certain turnover, but production is still flexible when the product is over the top and markets are slowly shrinking. Manufacturing capacity can be matched to meet product demand in a number of discrete steps. Applying the Equiplet manufacturing strategy will also limit overcapacity on the production floor (see table 1). After reaching the top of sales,

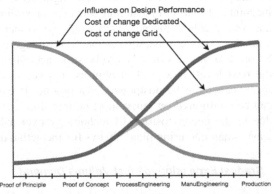

Fig. 3. Invested capital leads to increases of the cost for design changes of the product. Typically management involvement increases when facing the Process Engineering stage (halfway the graph). Influences to the Design Performance has drastically reduced.

production can be reduced using a limited number of production lines. The other manufacturing lines will be reconfigured to support a quick growing new product. In this sense it is constantly possible to tune the actual production to the exact demand.

4.3 Grid Agility

The manufacturing agility when applying grid manufacturing is improved by two main characteristics:

A. *The Equiplets are reconfigurable, having all advantages of reconfigurable equipment;*
B. *The Grid structure is used to create form-fit production facility. The higher number of systems lead to fairly accurate adjustment to market demand (see table 1).*

Ad A. Reconfigurable equipment combine productivity with acceptable cost in various volume demands [31]. Reconfigurable machines however are not the holy grail of manufacturing

$$L = \frac{1}{n} \cdot 100\%$$

No of parallel production platforms n	Max overcapacity L
3	33%
4	25%
5	20%
6	17%

Table 1: Overcapacity decreases when the number of parallel production systems grows

automation, but a efficient means for manufacturing products that serve a dynamic market and suffer from short life cycles (i.e. high end electronics, cellphones, PDA, gaming platforms). Once configured reconfigurable equipment acts as a dedicated machine for some amount of time.

Ad B. Manufacturing in Grids will add flexibility, because a larger number of machines will be divided over a number of products to be manufactured. How this works is shown in the next example (table 2):

In February 2010 a new product (I) will be launched that requires two processing operations of type A, three operations of type B and two operations of type C. In February 2010 a total number of eight Equiplets will take care of the production. Two of them will be configured as operation A, three as operation B another two as operation C (2A-3B-2C). A half-year later, market demand has increased, the production scheme now has become 4A-6B-5C. The same month a second product is launched (II). Again some Equiplets are configured to meet production demand. February 2011, the first reconfiguring action takes place. Two Equiplets, configured for operation A, are converted to operation D. Half a year later all Equiplets type A, and one of type B, have been reconfigured tot D. And so on.

Table 2. Example of dynamic configuration of a manufacturing grid. Production is expanded to roughly fifty Equiplets in 18 months to meet manufacturing demand for three products.

	February 2010	August 2010	December 2010	February 2011	August 2011
Product I	2 Type A 3 Type B 2 Type C	4 Type A 6 Type B 5 Type C	5 Type A 7 Type B 7 Type C	3 Type A 4 Type B 3 Type C	
Product II		3 Type B 5 Type C	7 Type B 11 Type C	8 Type B 12 Type C	6 Type B 8 Type C
Product III			1 Type B 1 Type C 1 Type D	4 Type B 4 Type C 5 Type D	9 Type B 10 Type C 12 Type D
Total No of Equiplets	2 Type A 3 Type B 2 Type C	4 Type A 9 Type B 10 Type C	5 Type A 15 Type B 19 Type C 1 Type D	3 Type A 16 Type B 19 Type C 5 Type D	15 Type B 18 Type C 12 Type D
	Total 8	*Total 23*	*Total 40*	*Total 43*	*Total 45*

In this case the factory floor grows to a number of roughly 50 Equiplets that are continuously converted to new production operations to meet the manufacturing demand for each actual product (A to D). Every now and then a machine is converted to another operation, causing minimal downtime, since the rest of the grid continues production. Machines are not necessarily logistically coupled to physical production lines, which is usually not interesting for the micro domain [16]. Adding logistical systems, conveyors, tray handlers etc, could however be economically feasible, especially near completion of the product when physical product volumes increase.

The example shows the excellent flexibility of the manufacturing grid, where transformation from one to another product can take place with minimal downtime. After initial configuration, the advantages get bigger when the reconfiguration takes place, since from that moment on, reuse of technology and investments contributes to more efficient application of resources. Again, the need for an effective DFA strategy should be emphasised. Since quick launch of production, as well as successful reconfiguration, only can succeed in close cooperation with R&D people.

4.4 Cost Modelling

Though in the past some cost modelling on micro production cells has been performed [16, 28, 29] there is no extensive literature on this topic available. Some first facts from literature:

- There is little or no need to invest in the automation of logistics. In view of weight and size it will be more cost effective to have operator bring parts to the machines in containers;
- Components need to be kept in batches, like trays, wafers or SMIF boxes to maintain orientation of products. This suppresses the need for realignment in every station;
- It may be expected that cost models will lead to new design rules for the set-up of micro factories;
- Cycle time remains an important factor, since its reduction directly leads to less equipment on the factory floor;
- Floor space could be a bottleneck. Dedicated equipment with a high engineering level is efficient on floor space. Microsystems manufacturing regularly take place under clean room environments, which lead to high cost of floor space. Though the footprint of the Equiplets will be small, a large number of them will be less efficient in comparison to dedicated machines.

5 Conclusions

The application of dedicated equipment causes disruptions during transformation of the product from R&D to the manufacturing phase. To avoid the disruptions during ramp-up, applied equipment is preferably to remain unchanged during the transition from R&D to full production. This would be the case if assembly tools during R&D stage, or at least the assembly principle and the conditions under which they are performed, could be reused for semi automated production and later eventually even automated production.

Using a DFA strategy, manufacturing complexity will be disentangled into basic assembly operations. These operations can be performed by relatively simple manufacturing platforms. By placing a larger number these platforms in a grid, a flexible manufacturing solution can be realised. The universal assembly platforms are called Equiplets, compact and low-cost manufacturing platforms that can be reconfigured to a broad number of applications by adding product dependent tools. These tools are very similar like the tools that are used during R&D and for the manufacturing of pre-production series. In this way existing R&D equipment is gradually upgradable, in a true re-configurable sense, investments in equipment are reduced significantly.

Especially when more agile manufacturing strategies are needed in an environment in which dealing with change becomes the central point, Equiplets can offer truly reconfigurable manufacturing, adaptable in small discrete steps, accurately matched to actual market demand.

Benchmarking Equiplet production has shown reduced time to market and a smooth transition from R&D to Manufacturing. When higher production volumes are needed, more systems can be placed in parallel to meet the manufacturing demand.

Costs of product design changes in the later stage of industrialisation have been reduced due to the modular production in grids, which allows the final design freeze to be postponed as late as possible. The need for invested capital is also pushed backwards accordingly.

The agility of the grid is unmatched. Reconfiguring machines can be done at a low level, reconfiguration of the Equiplets slows down production only partially. Total production shutdown, standard consequence in case of dedicated production lines, is prevented.

6 Future Research

Future research will address a cost model that can be used to estimate industrialisation cost of Equiplet-Grid manufacturing in comparison to dedicated and flexible manufacturing systems.

Secondly the contribution of modern (agent based) controller architectures will be investigated to see if self learning capabilities can further increase the ease of use of Grid Manufacturing with Equiplets, preferably in the most early R&D stage.

References

[1] Kear, F.W.: Hybrid assemblies and multichip modules. Dekker, New York (1993)
[2] Tichem, Tanase: A DFA Framework for Hybrid Microsystems. In: IPAS 2008, Chamonix (2008)
[3] Moore, G.A.: Inside the Tornado: Strategies for Developing, Leveraging, and Surviving Hypergrowth Markets: Collins Business (1995)
[4] Schünemann, Grimme, Irion, Schlenker, Stock, Schäfer, Rothmaier: Automatisierte Montage und Justage von Mikrokomponenten. wt Werkstattstechnik 89, 159–163 (1999)
[5] Grosser, V., Reichl, H., Kergel, H., Schünemann, M.: A fabrication Framework for Modular Microsystems. MST News, 4–8 (2000)
[6] EUPASS, EUPASS External Progress Report, in EUPASS External Newsletter (2008)
[7] Puik, E.: A Lower Threshold to the World by Combining a Modular Design Approach and a Matching Production Framework. In: COMS 2002 Ypsilanti, USA: Mancef (2002)
[8] Gutierrez, A.: MEMS/MST Fabrication Technology based on Microbricks: A Strategy for Industry Growth. MST News 1(99), 4–8 (1999)
[9] Becker: Micro- and Nanotechnologies for Advanced Packaging. MstNews 3(05), VDI/VDE/IT (2005)
[10] Heeren, Salomon, El-Fatatry, A.: Worldwide Services and Infrastructure for Microproduction. MstNews 2(04), VDI/VDE/IT (2004)

[11] Gunasekaran, A., Correa: Agile Manufacturing as the 21st Century Strategy for Improving Manufacturing Competitiveness (2001)

[12] Goldhar, Jelinek: Plan for Economies of Scope. Harvard Business Review, 141–148 (1983)

[13] Stecke, Raman: Production Flexibilities and their Impact on Manufacturing Strategy. Graduate school of business administration, U. o. Michigan (ed.)

[14] Salomon, Richardson: Design for Micro & Nano Manufacture. MstNews. 3(05) (2005)

[15] Hollis, Rizzi: Agile Assembly Architecture: A Platform Technology. In: ASPE American Society for Precision Engineering Orlando, Florida, USA (2004)

[16] Koelemeijer-Chollet: Cost Modelling of Microassembly. In: International Precision Assembly Seminar 2003 Bad Hofgastein, Austria (2003)

[17] Mardanov, Seyfried, Fatikow: An Automated Assembly System for a Microassembly Station. Computers in Industry 38, 93–102 (1999)

[18] Hollis, Gowdy: Miniature Factories for Precision Assembly. In: International Workshop on Micro Factories, Tsukuba, Japan (1998)

[19] Yang, Gaines, Nelson: A Flexible Experimental Workcell for Efficient and Reliable Wafer-Level 3D Microassembly. In: International Conference on Robotics & Automation, Seoul, Korea (2001)

[20] Fahlbush, F., Seyfried, Buerkle: Flexible Microrobotic System MINIMAN: Design, Actuation Principle and Control. In: Advanced Intelligent Mechatronics, IEEE/ASME International Conference Atlanta, GA, USA (1999)

[21] Zhou, Aurelian, Chang, Corral, Koivo: Microassembly System with Controlled Environment. In: Microrobotics and microassembly, Newton, ETATS-UNIS (29/10/2001), vol. 4568 Conference No3, pp. 252–260 (2001)

[22] Kim, Kang, Kim, Park, Park: Flexible Microassembly System on Hybrid Manipulation Scheme. The International Journal of Advanced Manufacturing Technology 28, 379–386 (2005)

[23] Gaugel, Bengel, Malthan: Building a Mini-Assembly System from a Technology Construction Kit. Assembly Automation 24, 43–48 (2004)

[24] Tanaka: Development of Desktop Machining Microfactory. Riken Rev. J0877A 34, 46–49 (2001)

[25] Lauwers, Edmondson, Hollis: Progress in Agile Assembly: Minifactory Couriers Based on Free-Roaming Planar Motors. In: 4th International Workshop on Microfactories, Shanghai, China, pp. 7–10 (2004)

[26] Clévy, Hubert, Chaillet: Micromanipulation and Micro-Assembly Systems. In: IEEE/RAS International Advanced Robotics Programm, IARP 2006 Paris, France (2006)

[27] Cecil, Gobinath: Development of a Virtual and Physical Work Cell to Assemble Micro Devices. In: 14th International Conference on Flexible Automation and Intelligent Manufacturing. Robotics and Computer-Integrated Manufacturing, pp. 431–441 (2004)

[28] Koelemeijer-Chollet: Economical Justification of Flexible Microassembly Cells. In: International Symposium on Assembly and Task Planning Besancon (2003)

[29] Koelemeijer-Chollet: A Flexible Microassembly Cell for Smal and Medium Sized Batches. In: Proceedings of 33rd International Symposium on Robotics (2002)

[30] Roussel, P.A., Saad, K.N., Erickson, T.J.: Third Generation R & D: Managing the Link to Corporate Strategy. Harvard Business Press (1991)

[31] Dashchenko (ed.), Koren: Reconfigurable Manufacturing Systems and Transformable Factories. General RMS Characteristics. Comparison with Dedicated and Flexible Systems. Springer, Heidelberg (2006)

Development of a Reconfigurable Fixture for the Automated Assembly and Disassembly of High Pressure Rotors for Rolls-Royce Aero Engines

Thomas Papastathis[1], Marco Ryll[1], Stuart Bone[2], and Svetan Ratchev[1]

[1] School of Mechanical, Materials and Manufacturing Engineering,
The University of Nottingham
{epxtp,epxmr3,svetan.ratchev}@nottingham.ac.uk
[2] Rolls-Royce plc, Technology Acquisition Department
stuart.bone@rolls-royce.co.uk

Abstract. This paper describes a novel fixturing system for the automated assembly and disassembly of aero engine components. The proposed system has the ability to automatically reconfigure for a number of different parts and allows the precise control and monitoring of its clamping positions and exerted clamping forces. Aspects of the control software of this system, like the graphical user interface (GUI) and the relational data base system containing postprocess and configuration data, are described.

1 Introduction

The manufacturing of aero engines and its sub-modules is traditionally characterised by rigorous product quality requirements and tight tolerances. Additionally, as a result of the global competition, increased product diversity is fast becoming a requirement for the aerospace sector too. At the same time, aero engine manufacturers, like Rolls-Royce, are adopting the Product Service System (PSS) paradigm. Because of the latter, manufacturers are not only working towards reducing lead times, but also minimising service and repair costs and efforts. For this reason, modern aero-assembly processes need to be upgraded by automated and rapidly reconfigurable manufacturing solutions in their production environment. Fixtures are a vital part of this environment which are devices designed to locate, support and hold workpieces in a desired position for an operation to be performed. In the past decades, industry and researchers have shown an increased interest in fixtures and their effect on the manufacturing outcome. In particular, a number of research work has addressed the issues of reconfigurable fixturing systems. Al-Habaibeh *et al.* [1] proposed a reconfigurable fixture based on an array of locator pins. Offering increased flexibility and distributed clamping load, these fixtures have been trialled on small and flexible aero components such as Nozzle Guide Veins (NGVs). However, locating precision issues and limited accessibility have limited their application on the shop floor. Chakraborty *et al.* [2, 3] have presented a flexible fixturing system for automobile engine cylinder heads which is able to identify and reduce the positioning error, thus increasing dimensional accuracy of the final product. Other reconfigurable fixturing

S. Ratchev (Ed.): IPAS 2010, IFIP AICT 315, pp. 283–289, 2010.
© IFIP International Federation for Information Processing 2010

solutions have been presented in [4-6]. However, these solutions have not been applied in industrial application.

Recently, in an attempt to further decrease workpiece deformation and thus improve the final product quality, a number of researchers have tried to actively adapt the clamping forces of the fixture to external influences [7]. Tao *et al.* [8, 9] have experimentally proven that clamping loads and workpiece deformation can be drastically reduced if the clamping forces are maintained to an ideal value during the machining process. Liao *et al.* [10] further investigated on the effects of clamping forces on the surface quality of machined automobile engine blocks. It was shown that clamping forces affect significantly the surface quality of the machined part, especially as the workpiece rigidity reduces. Despite the advantages of adaptive fixturing systems, this type of fixtures has seen limited transfer to industrial environments.

The review of the reported research indicates that a number of approaches exist for either reconfigurable fixturing or adaptive fixturing. However, the current solutions do not combine these two aspects into an integrated approach, the benefits of which could be significant, especially for industrial application. Additionally, the proposed solutions for adaptive fixtures appear to concentrate mainly on machining operations and are not adequately addressing the needs of aero assembly operations.

In an attempt to overcome these gaps and meet the future challenges of aero engine assembly within Rolls Royce plc., this paper reports on the ongoing development of a fixturing system for aero assembly components that is both reconfigurable and adaptive. The remaining paper is organized as follows. Section two describes the current process and future challenges for Rolls Royce. This is followed by an electro-mechanical system overview of the fixture in section three, while section 4 discusses aspects of the software architecture.

2 Current Process Description

The proposed fixturing system will be used for the assembly of the various stages of the High Pressure Compressor (HPC). This component sits in the heart of the jet engine (Fig. 1) and comprises various cylindrical stages, with diameters ranging from a few hundreds of millimetres to three thousand millimetres depending on the engine

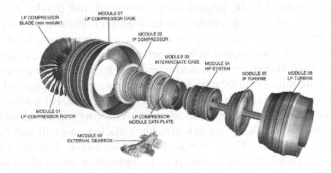

Fig. 1. Modular break-down of a Rolls-Royce Trent Engine

model. Rotating at very high speeds and being exposed to enormous temperatures during operation, the HPC must conform to strict quality requirements regarding vibration, geometrical run-out and swash. On top of that, geometrical differences between the parts of the different engine variants demand fixture set up changes that elongate the overall lead time. These challenges in the assembly process of the HPC, are also present during disassembly for rework, service and repair purposes and subsequent re-assembly.

Currently, the assembly and disassembly process of HPCs is mainly manual, involving a series of activities performed by human operators. As an essential part of the operation, a hydraulic fixture is utilised to clamp the stages in the desired position for a predetermined amount of time. The latter is entirely operator controlled and its physical reconfiguration to accommodate the different part sizes, is carried out manually. Although this fixturing solution fully satisfies the present performance and process throughput demand, envisaged developments within Rolls-Royce predict the necessity to reduce lead times and further increase the repeatability of the fixturing outcome. Nevertheless, the current process could face three main challenges in the future. Firstly, the time-to-reconfigure and assembly/dis-assembly process cycle times characterising the current fixturing system could prove to be a bottleneck when looking to reduce production and servicing times. Manual reconfiguration could inhibit the necessary fast response when a setup change-over is requested. Speeding up the assembly and dis-assembly processes would also help reduce lead-times. Secondly, the currently acceptable percentage of parts failing the quality tests and demanding rework needs to be reduced as more parts are to be manufactured per year. This will keep repair and rework costs to a minimum. For this reason, a more repeatable process is necessary, thus the part of the fixturing cycle that is dedicated to manual labour needs to be reduced and the process parameters optimised and more tightly controlled. Finally, the ability to monitor clamping forces and generate automatic post-production reports, which is not available in the current fixturing system, could help identify performance variations and assist in adapting the process more effectively to future requirements.

3 Design Overview of Fixturing System

A simplified overview of the proposed fixture is shown in figure 2. The proposed fixture, developed in cooperation with ETN Aviation GmbH, comprises of 6 fixturing elements, i.e. three clamps and three support elements. The upper three elements are used to provide clamping and separating forces to the upper stages, whilst the three support elements are used to simply locate and hold the lower stages during the disassembly process. In total the fixture deploys three direct current (DC) servomotor actuators and nine off-the-shelf step motor rotary actuators on a circular frame to allow repositioning of the clamps in radial and vertical direction.

These actuators grant the fixture the ability to automatically reconfigure for a number of different cylindrical parts. In detail, three DC ToX ElectricDrive EPMK electromechanical actuators are used in a vertical orientation to provide the clamps with the capability to clamp the stages together during the assembly process with a specified amount of force and over a predefined period of time. Additionally, in case

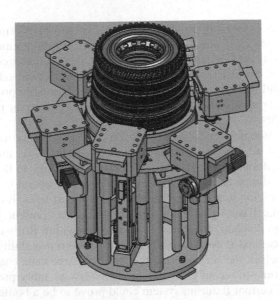

Fig. 2. Simplified design overview of the fixturing system

of servicing and repair activities, they are used to disassemble the HPC by pushing the stages apart. The actuators are controlled by three separate ToX L9400 X-Line controllers. Communication of the controllers is established via ProfiBus, ensuring tight control of actuation times to achieve synchronised clamping. Individual actuation of each clamp is also possible. The actuators are equipped with force feedback through strain gauges and position feedback via encoders. These provide the necessary information for maintaining stable clamping forces over the process cycle and for reporting purposes at the end of the cycle. Feedback signals are also transferred to a PC via ProfiBus.

Three NanoTech AD8918 stepper motors driving an equal number of ball screws are used to actuate vertically the support elements. Finally, six stepper motor based linear actuators are used to achieve radial movement of both the clamps and the support elements. All nine stepper motor are powered via an equal number of SMCI47 stepper drivers. Communication of the PC to these drivers is achieved through an Advantech PCI-1758U I/O card. An array of ten Balluf BESM08EHPSC15B-S04 proximity sensors is used to reference the system at startup.

The use of electromechanical actuation and the combination of the previously mentioned actuator types produces a compact and clean solution with only marginally higher cost than an equivalent hydraulic solution. Nevertheless, this design gives the extra advantage of allowing maximum control over process parameters such as actuating forces, clamp actuation sequence and clamp positions. The latter is especially important for the automatic reconfiguration of the fixture for different part sizes.

4   Overview on the System Software

Operators can control the operation of the fixture through a PC-based control environment. The latter consists mainly of three components: (1) a data base system storing all relevant process parameters for each workpiece and for both the assembly and disassembly; (2) a graphical user interface guiding the operator through reconfiguration sequence and (3) a series of communication interfaces to the controllers of the actuators on which the force profiles of the clamps are stored. When a new workpiece is detected, the control software can retrieve the adequate process parameters from the data base and subsequently trigger the related clamping profiles on the actuators. For this purpose, the software communicates with the actuator controllers of the fixture via a ProfiBus interface and digital I/O connections as described in the previous section. During the process the software monitors the online force and position feedback from the clamps and stores all relevant data permanently on the data base. This information can be used to generate post-production reports, which is realised with the LabView Diadem Reporting functionally. An overview on the entire system is provided in figure 3.

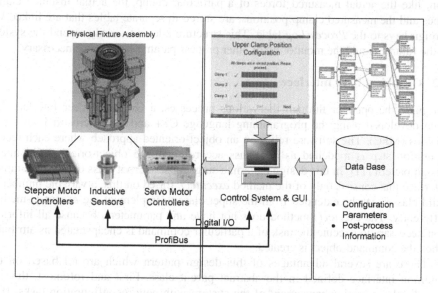

Fig. 3. Overview of the software and communication system

In the following sections, important aspects of the data base structure and the user interface are briefly described. The communication interfaces and routines on the actuator controllers shall be omitted here for brevity.

4.1 Data Base Design and Implementation

An entity-relationship model (ERM) has been developed to represent the relations and characteristics of the various engine types, processes, fixture configuration parameters and post-process data. This model has been implemented with the Microsoft Access database system. Key to this model is the table *EngineTypeDef*, which holds information about the various engine types, including an identifier and a name. Linked to this table via foreign keys, the table *StageTypeDef* contains information about the structure of an engine type while the available processes for a certain engine type are defined in the table *ProcessDef*. Each process is defined by a numerical id, a name and a link to the engine type table (foreign key). This allows the grouping of the required fixture configuration parameters for a particular engine/process pair in the table *ClampConfiguration*, which contains the clamping positions, home positions, desired force profiles etc. Each record in this table is uniquely identified by a so-called primary key, consisting of a numerical clamp id and the process id, which links back to the *ProcessDef* table. Consequently, the data model can be easily adapted to future fixture designs because an arbitrary number of clamps can be defined for each process. When the system is in operation, a record for the particular engine and process is created in the data base tables *EngineLog* and *ProcessLog*. The former contains a serial number and is linked to the *EngineTypeDef* table. The latter is linked to the *ProcessDef* table and *EngineLog* and holds global information of the particular process such as start time, end time and a link to the operator. More detailed process information, like the actual measured forces of a particular clamp, the actual installed clamp types and the measured clamp positions, are stored in separate tables that are linked via foreign keys to the *ProcessLog* table. This structure allows to easily extend the system in the future, should the monitoring of other process parameters become necessary.

4.2 Graphical User Interface

To guide the operator through the various processes, a Graphical User Interface has been developed using the programming language C++ and the Microsoft Foundation Classes (MFC). The software relies on an object-oriented approach where each reconfiguration step is modelled as one class according to the object-oriented Command design pattern [11]. In this design pattern, a single abstract root class defines a common interface that consists only of the method *execute()*. Based on this, a variable number of child classes can be extended from the super class to implement the *execute()*-method differently. The *execute()*-method does not have any parameters, because all information necessary to execute the task of a particular command is encapsulated as attributes when the command-object is created.

There are several advantages of this design pattern which are all based on the common interface defined in the abstract parent class. First and foremost, this approach allows easy enhancement of the system with new reconfiguration tasks. If in the future the necessity of further reconfiguration steps evolves, the class hierarchy can easily be extended without affecting the rest of the system. Secondly, decomposing the entire fixture reconfiguration into atomic steps modelled as single objects significantly simplifies the complexity for control software. In more detail, due to the common interface, the execution of the entire reconfiguration sequence is reduced to simple calls of the *execute()*-method of each command.

5 Conclusions

In this paper, hardware and software aspects of a new fixturing system for the automated assembly of aero-engines has been described. The system is addressing three major requirements of future aero-assembly, namely the reduction of reconfiguration times, higher repeatability due to better process control and the ability to monitor process parameters and generate post-process reports. The fixturing hardware consists of three DC-servomotor actuators that generate the clamping forces and a series of stepper motors responsible for the positioning. Key to the control software is a graphical user interface, that guides the operator through the reconfiguration sequences and a relational data base system, holding important process and configuration data.

Acknowledgments. The reported research is conducted as part of the ongoing European Commission FP6 funded integrated project (FP6-2004-NMP-NI-4) - AFFIX "Aligning, Holding and Fixing Flexible and Difficult to Handle Components". The financial support of the EU and the contributions by the project consortium members are gratefully acknowledged.

References

1. Al-Habaibeh, A., Gindy, N., Parkin, R.M.: Experimental Design and Investigation of a pin-type reconfigurable clamping system for manufacturing aerospace components. Journal of Engineering and Manufacture - Proceedings of the Institution for Mechanical Engineers Part B 217, 1771–1777 (2003)
2. Chakraborty, D., De Meter, E.C., Szuba, P.S.: Part location algorithms for an intelligent fixturing system part 1: system description and algorithm development. Journal of Manufacturing Systems 20, 124–134 (2001)
3. Chakraborty, D., De Meter, E.C., Szuba, P.S.: Part location algorithms for an Intelligent Fixturing System Part 2: algorithm testing and evaluation. Journal of Manufacturing Systems 20, 135–148 (2001)
4. Chan, K., Benhabib, B., Dai, M.: A reconfigurable fixturing system for robotic assemble. Journal of Manufacturing Systems 9, 206–221 (1990)
5. Du, H., Lin, G.C.I.: Development of an automated flexible fixture for planar objects. Robotics and Computer-Integrated Manufacturing 14, 173 (1998)
6. Youcef-Toumi, K., Buitrago, J.H.: Design and implementation of robot-operated adaptable and modular fixtures. Robotics & Computer-Integrated Manufacturing 5, 343–356 (1989)
7. Nee, A.Y.C., Tao, Z.J., Kumar, A.S.: An Advanced Treatise on Fixture Design and Planning. World Scientific Publishing, Singapore (2004)
8. Tao, Z.J., Kumar, A.S., Nee, A.Y.C.: Automatic generation of dynamic clamping forces for machining fixtures. International Journal of Production Research 37, 2755 (1999)
9. Tao, Z.J., et al.: Modelling and experimental investigation of a sensor-integrated workpiece-fixture system. International Journal of Computer Applications in Technology 10, 236–250 (1997)
10. Liao, Y.G., Hu, S.J.: An integrated model of a fixture-workpiece system for surface quality prediction. International Journal of Advanced Manufacturing Technology 17, 810–818 (2001)
11. Gamma, E., et al.: Design Patterns. Elements of Reusable Object-Oriented Software. Addison-Wesley Longman, Amsterdam (1995)

Chapter 9

Micro-factory

Architectures and Interfaces for
a Micro Factory Concept

Niko Siltala, Riku Heikkilä, Asser Vuola, and Reijo Tuokko

Tampere University of Technology, Department of Production Engineering
Korkeakoulunkatu 6, 33101 Tampere, Finland
{niko.siltala,riku.heikkila,asser.vuola,reijo.tuokko}@tut.fi
http://www.tut.fi/tte

Abstract. So far the desktop manufacturing is mainly done as islands of process modules or in some seldom cases the desktop factory is created in form of manufacturing line. Tampere University of Technology has been working on such desktop factory concepts for years and come out a microFactory concept (TUT-μF). The paper discusses architectural aspects and proposes some solutions for them. It specifies also two main mechatronic interfaces used for such modular desktop factories - 1) the cell to cell interface and 2) cell internal process module interface. Main parts of the specifications are represented. These can be utilised for building the desktop production line from easily integrated modules.

Keywords: desktop factory, micro factory, module interfaces.

1 Introduction

This paper will present the recent development and documentation of the architectures and interfaces for a micro factory concept developed at Tampere University of Technology (TUT). The concept is called TUT-Microfactory© shortened here as TUT-μF. The concept has long roots throughout the series of national projects starting in 2000 from project TOMI (Towards Mini and Micro Assembly Factories). Since then the concept has gone through several evolutions and grown more mature as a concept. [1, 2]

The different mini and micro factory concepts have been presented since early 90s [3]. However, the research is typically focusing on single machines and not that much on integration of stand-alone processes and machines into larger manufacturing entities like production lines with integrated material logistics. Motivation to research and development of micro and desktop manufacturing and factories leans on technological, business and sustainability reasons. The first because there exist a need for manufacturing and assembly of high-precision, miniature products; and new innovations and technical solutions offer today smaller sized components for building the systems. The second, because there exist a need to postpone the product customisation as late as possible and closer to the customer; and possibility to utilise new business models. The latter because arrival of new requirements for space savings, energy consumption and utilisation of resources on factory floor. Modular systems built from reconfigurable and re-usable modules support as well the sustainability aspect.

S. Ratchev (Ed.): IPAS 2010, IFIP AICT 315, pp. 293–300, 2010.

Opening the architectures and interfaces the TUT-μF concept will be available for others to join and utilise the concept and also to provide the new compatible process modules. System integration will become easier as the shared interfaces (both mechatronic and communication) are used between modules from different origins.

2 Method

In many cases the mini, micro or desktop factory concept or application is working like an isolated island, which is served by a human operator and assisting him or her by performing tasks which are difficult for a human. Integration of the material flow between different stations or integration on the control side is the typically missing. The new option for micro factories is an integrated assembly or manufacturing system where workstations or cells are aggregated into a line or even bigger factory system. The cells itself can also be aggregated from modules with standardised interfaces. The proposed TUT-μF concept is based on the latter integrated solution having as one objective to provide an architecture where the process specific modules can be composed freely into different layouts with minimum limitations; modularity, reconfiguration and re-use are well presented; each module is self-contained with their controls; etc.

3 Architectures

3.1 Cell

The outer dimensions of the base module (i.a. cell frame) used in the TUT-μF are 200 x 300 x 230 mm (Width x Depth x Height). The cell size is extendable in height as the process modules can be stacked on top of the base module. The depth of the base module is allowed to vary within the given range depending on the need of cell manufacturer and implemented processes. The line width is the most rigid dimension off the base module. The inside work space is 180 x 180 x 180 mm. This offers small modules easily movable by human means, but still sufficient workspace volume needed for assembling typical handheld electronic products and other mini or micro-sized products. Cell communication architecture is presented in [6].

The base module can have other supplementary features implemented like the clean room or temperature controlled environment, frequency (wavelength) blockage shielding e.g. in case of laser processes.

The cell itself and the processes it is capable to perform are specified with the Emplacement concept and Blue Print files [5]. This is offering a novel method for electronically describe the features and capabilities of a module. This information can be utilised during the system design and use [4].

Cell Internal Architecture. One main cell internal interface is the process module interface on top of the cell, which can be repeated allowing a stacking structure. The electrical supply and communication through ethernet is available. Modules are intended to be self-contained and independent modules with their own controls. The base plate of the base module can be used to connect transport and process related sub-modules. The matrix of attachment locations is provided through thread holes.

3.2 Line Layout

The base modules can be relatively freely and easily set up on different line layouts. The layout can be line with branches, even some loops are allowed. The Fig. 1 illustrates one possible layout example including a carrier based transport system.

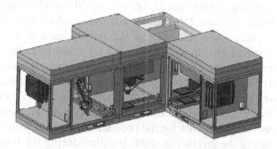

Fig. 1. Example of TUT-Microfactory layout [1]. Represented are four cells with different manipulators and conveyor system connected together.

Even the maximum freedom for combining cells together is sought, some rules still apply. The Fig. 2 illustrates some of these rules. The arrow inside the cell points the supposed flow direction of the main product (from left to right). The male and female connectors are differentiated by the shape and correspondingly the purpose of the connection with colours. The basic rule is that male and female connectors with the same colour can be connected together. As one can recognise, the mechanical connection is possible in some other configurations, but the purpose of connectors will then mismatch which leads to a critical signal failure. However, some techniques could be used in order to get such kind of layouts accepted e.g. by using a cross-wiring adapter or change-over switch for signals.

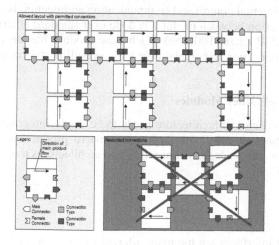

Fig. 2. Allowed and restricted cell connections within the TUT-µF architecture

3.3 Transport

Three different forms of transport are proposed: pallet based, product on conveyor and product on air. The first is traditional pallet based approach using carriers to transport the product and other materials between workstations. Carrier can be based on e.g. DIN 32561 [7]. Simple belt conveyors or other means can be used for making the transport. The advantages are stable support for differently shaped and sized parts travelling through the system. Drawbacks are the product specific caves or other mountings made to pallet and returning of the pallets back to the beginning of the system. The carrier based system within the TUT-µF concept is discussed in [3].

In the second approach the product is transported without a specific carrier from a workstation to next on a flat belt conveyor. This approach is suitable for the first case so it can also move the pallets. Here the advantage is the universality of the solution. There is no need for product specific fixtures on the pallets. However the main drawback is the additional requirements for the product. It needs to have size large enough, flat bottom maintaining the product in stable position during the transport between the workstations and manipulations, and surface quality for the product should not be very high (as the pallet is not protecting the product). In many cases it is required to have a separate fixture inside the cell for supporting the product during the assembly.

In the third approach the product is passed on air with the help of manipulators between the parallel workstations. This would require means to grip the part and transfer it between the processing places. The manipulator making the assembly can be utilised for the transfers, if the range is large enough.

In all cases the by-pass of the cell must exist. In cases when the workstation gets broken and production needs to continue or there are multiple stations performing the same process task, the by-pass gets mandatory. The performance of the overall system depends on the way the by-pass is implemented.

3.4 Controls and Communication

Ref. [3] discusses the possible control approach followed by the concept. One instance of the cell internal control and communication architecture is opened in [6].

The communication between modules is done over standard ethernet protocols like TCP/IP and UDP/IP (max. 100Mbit). Each base module has an embedded switch distributing the connection to neighbouring cells as well to the control modules composing the cell.

3.5 Integration of Other Modules

The TUT-µF approach and architecture needs to be easily integrated with other kind of systems like lines or cells with larger size (macro production) or with different architecture. These cases adapters connecting two architectures together are used.

4 Results

4.1 Interfaces

The standardised interfaces are the main offering of the proposed concept even they are yet only "company specific standards". These are introduced and discussed in details on the following chapters.

Cell-to-Cell. The cell to cell interface for TUT-µF is specified in [9]. The main part of the interface is represented in the Fig. 3, which shows the dimensional drawing of the interface. It shows the details of the mechatronic interface including the intended features, dimensions, connectors used. In addition to mechanical information the electrical signals, pin assignments, communication channels, origin, etc. are specified in the document.

The features available in the specification are: Opening for the product to pass through. At minimum 180 x 180 mm opening shall be available. In the practice the application may use smaller area; Fastening and fixation (1A & 1G in Fig. 3). Hinges and plates can be used to fixating cells together; Positioning function (1B & 1F). The pins and holes or bushes are used to accurately locate the mating interfaces. The pin (1F) serves as the origin of the interface.; Electrical connection and communication (1C & 1E). The electrical supply of 24 VDC is supplied through the larger pins. Ethernet (max.100Mbit) and handshake signalling for the product exchange between cells are using the smaller signal pins; Pneumatics distribution line in (1D).

The connectors are on the same base line ([A] in Fig. 3), from which the cell floor at processing area is located 20 mm above and cell bottom 10 mm below. The cell floor is allowed freely to go below the defined, have openings for manufacturing processes, etc.

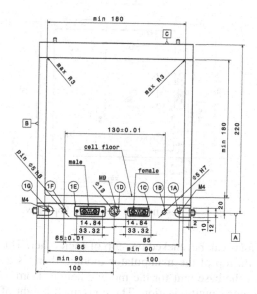

Fig. 3. Cell-to-cell interface. View at front.

Transport. Transport interface for the main product flow is specified in [9]. There are specified three different options for cell to cell transport of products on the main production flow. Two of them are conveyor based approaches and third is through air i.e. manipulator is taking care of the transport over the cell boundary.

The conveyor based approaches define where the conveyor shall be located, how the used conveyor width is designated. The main principles are that in the option A the conveyor is located back half of the workspace leaving the front part for assembly and feeding actions. In the option B the conveyor is located at the centre line of the side interface of the cell i.e. going through at the centre of the workspace of the cell from the left to right.

Process Module. The process module interface for TUT-μF is specified in [10]. The main part of the interface is represented in the Fig. 4. It shows the dimensional drawing of the interface with the details of the mechatronic interface including the intended features, dimensions and connectors used.

The interface specification [10] defines the features like: Minimum available opening for the process modules above is defined (180 x 180 mm); Orientation feature with the use of pins (2B) in Fig. 4, which defines the orientation of the interface; Interface origin is located concentric with the left side pin of (2B); Fastening of the process module with bolts having M4 thread; Electrical supply and communication. The connection contains power for 24VDC and ethernet communication. Additional communication channels can be passed through parallel with ethernet. As option are defined pin assignments for busses like USB, industrial fieldbuses, etc. The specification document defines the connectors, electrical signals, pin assignments, etc.

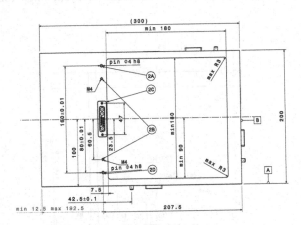

Fig. 4. Process module interface. Top view.

The process modules can be stacked one top of each other. This will be very useful in cases e.g. when on top of the manipulator module there is a vision module. The specification defines the base unit for the module height, from which the real implemented module is integer multiplication. The consumed height of the module will be included into the module designation information. If stacking is not allowed for some reason it is also described by the designation.

Manipulator - End effector. TUT-μF concept utilises the ISO 29262 [8] for the interface between the manipulator and end effector.

Feeding. Ref. [3] discusses some feeding methods utilised in TUT-μF like tray and flexible feeding. The interface specification for the feeding is not yet stabilised and found its latest form within our concept for single feeders. One of proposed alternative is to use Fuji IP/QF series feeder interface. However, it will be sure that several different options will be finally available within the concept, as one single approach cannot be serving all different objectives like mechanical scalability of size (both component and feeder), feeding methods, etc.

In case the feeder is independent and self-contained module, it can be even connected through the cell interface (See Fig. 3)

4.2 Standardisation

The objective is extensive use of the existing standards. However as we are breaking new areas no standards yet exist. The field of micro factories is relatively new and this is one of the first attempts to standardising the interfaces used for such environments.

The paper mentions some standards, which are applied within the concept like [7, 8]. It also presents some specifications which are describing the interfaces used in the TUT-μF concept. These specifications are currently on the development phase and TUT will publish them through some channel later this year. One possible channel could be Evolvable Assembly Systems Environment (http://www.eas-env.org/).

5 Conclusions and Future

The paper presents the details of some key interfaces for the TUT-Microfactory (TUT-μF) concept, which are further developed and documented within our currently ongoing Mz-DTF project at Tampere University of Technology. By publishing the interfaces we invite others to join the development of the specifications and to utilise these in order to be able to realize multi-vendor micro and desktop factory systems based on components and modules compatible with each other as they all basis on the same shared definitions and architecture.

In future we are planning to extend the existing interfaces and adding new ones to the set of specifications in the TUT-μF concept. The solidification of our specifications through real standardisation will be also included in the future work.

References

1. Heikkilä, R.H., Karjalainen, I.T., Uusitalo, J.J., Vuola, A.S., Tuokko, R.O.: Possibilities of a Microfactory in the Assembly of Small Parts and Products – First Results of the M4-project. In: Proceedings of IEEE International Symposium on Assembly and Manufacturing (ISAM) 2007, Ann Arbor, Michigan, USA, July 22-25, pp. 166–171 (2007)
2. Heikkilä, R., Huttunen, A., Vuola, A., Tuokko, R.: A Microfactory Concept for Laser-Assisted Manufacturing of Personalized Implants. In: Proceedings of 6th International Workshop on Microfactories (IWMF 2008), Northwestern University, Evanston, Illinois, USA, October 5-7, pp. 77–80 (2008)
3. Jarvenpää, E., Heikkilä, R., Tuokko, R.: Logistic and Control Aspects for Flexible and Reactive Micro and Desktop Assembly at the Factory Level. In: Proceedings of 2009 IEEE International Symposium on Assembly and Manufacturing (IEEE ISAM 2009), Suwon, Korea, November 17-20, pp. 171–176 (2009)

4. Siltala, N., Tuokko, R.: Use of Electronic Module Descriptions for Modular and Reconfigurable Assembly Systems. In: Proceedings of 2009 IEEE International Symposium on Assembly and Manufacturing (IEEE ISAM 2009), Suwon, Korea, November 17-20, pp. 214–219 (2009)

5. Siltala, N., Tuokko, R.: Emplacement and Blue Print - Electronic Module Description supporting Evolvable Assembly Systems Design, Deployment and Execution. In: 6th International Conference on Digital Enterprise Technology (DET 2009), Hong Kong, December 14-16 (in press, 2009)

6. Vuola, A., Heikkilä, R., Prusi, T., Remes, M., Rokka, P., Siltala, N., Tuokko, R.: Miniaturization of Flexible Screwing Cell. In: Ratchev, S., Hauschild, M., (eds.): IPAS 2010, IFIP AICT 315, pp. 309–316 (2010)

7. DIN 32561: Production equipment for microsystems - Tray - Dimensions and tolerances. DIN standard (2003)

8. ISO/DIS 29262: Production equipment for micro-systems - Interface between endeffector and handling system. ISO Draft international standard (2009) (Supersedes DIN 32565 E) (February 2005)

9. TUT-μF: std-0001: Cell Interface. Draft specification (2010) (unpublished)

10. TUT-μF: std-0002: Process Module Interface. Draft specification(2010) (unpublished)

Desktop Micro Forming System for Micro Pattern on the Metal Substrate

Hye-Jin Lee[1], Jung-Han Song[1], Sol-Kil Oh[1], Kyoung-Tae Kim[1], Nak-Kyu Lee[1], Geun-An Lee[1], Hyoung-Wook Lee[2], and Andy Chu[3]

[1] Manufacturing Convergence R&D Department,
Korea Institute of Industrial Technology, Korea
[2] Dept. of Energy System Engineering, Chungju National University, Korea
[3] Space Solution Co., Ltd. Seoul, Korea
{naltl,jhsong,solkil,ktkim,nklee,galee}@kitech.re.kr,
hwlee@cjnu.ac.kr, crimson1@spacesolution.kr

Abstract. In this Research, the desktop micro forming manufacturing system has been developed. A micro forming system has been achieved in Japan and its developed micro press is limited to single forming process. To coincide with the purpose to be more practical, research and development is necessary about the press which the multi forming process is possible. Micro patterned metal components are used in so many precision engineering fields. This micro pattern plays an important part in the functional movement of precision module. This micro pattern on the metal component can be made by EDM(Electro Discharge Machining). But this EDM method has low productivity because EDM tools can be worn easily. If another manufacturing process is developed with high productivity, industries can product the competitive goods. So we research on the forming process and system to make micro functional pattern on the metal component.

Keywords: Micro Metal Forming, Forming Manufacturing System, Micro Thin Foil Valve, Multi Processing.

1 Introduction

The existing forming press uses a hydraulic actuator and high powered mechanical actuator (Fig.1-(a)), therefore occupying a large space because of its size. This type of system is inefficient for manufacturing micro size and precision products.

As forming components are small in size, forming equipment must also be small in size because the forming die and load must be small. The micro manufacturing press system is an ultra precision forming equipment the size of several micros to millimeters and precision of sub-micro to micrometer. This system can be applied to a micro factory system module that manufactures micro components using micro thin foil and bulk material. This micro forming manufacturing system (Fig.1-(b)) has the advantage of minimization in manipulating distance and working space. As equipment and tools become smaller in size, minute inertia force and high natural frequency can be obtained. Therefore, high precision forming performance can be obtained. This allows the factory to quickly provide the customer with goods because the manufacturing system and process are reduced. To construct a micro manufacturing system, many technologies are necessary such as high stiffness frame, high precision actuating part, structural analysis, high precision tools and system control.

S. Ratchev (Ed.): IPAS 2010, IFIP AICT 315, pp. 301–308, 2010.
© IFIP International Federation for Information Processing 2010

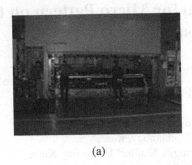

(a)   (b)

Fig. 1. (a) Existing forming press and (b) Micro forming press

Most of the research about microscopic sized forming is underway concentrating in forming on the polymer substrate. But the research on the micro forming equipment miniaturization and forming technology on the metal substrate are insufficient than the microscopic sized forming technology on the polymer substrate.

This research is about research on the modified desktop micro forming system to manufacture the micro pattern on the metal substrate and maximize the efficiency of micro forming process. This desktop micro forming system is developed in collaboration with KITECH (Korea Institute of Industrial Technology) and Space Solution Inc. The modified system is consists of gear type actuating module and micro compressive forming die-set. The forming results of micro herringbone pattern are described to verify the performance of developed system.

2   Related Research about Micro Manufacturing System

The micro machine project was the initial step in the micro manufacturing system. This project has become the leading research in micro manufacturing with national support by MITI (Economy, Trade and Industry Ministry). MITI has infused a total of two-hundred million dollars into this project over the past 10 years since 1991. In this project, AIST (National Institute of Advanced Industrial Science and Technology) developed a micro milling machine, micro lathe, micro press, component transport section, and assembly section on a flat table the size of 500mm x 700mm. AIST manufactured a miniature bearing module using this micro factory system. A picture of the micro press system is shown in Fig. 1b and the size of this micro press system is 111mm x 66mm x 170mm (length x width x height). This system is driven with a 100W AC serve motor and a maximum load capacity of 3kN. AIST manufactured a Φ1mm miniature bearing module cover part in a bending process using the micro press system. However, research has been restricted to the bending process and its practical efficiency is low.

TIT (Tokyo Institute of Technology) designed the micro punching press system as shown in Fig.2. A micro pump using a piezo is used for an actuator of the micro punching press system. TIT carried out a 300μm hole punching process on the 10μm thickness aluminum foil. The punching force is 1.85kN in this process. The pictures of the punch and the fabricated micro hole are shown in Fig.3.

Fig. 2. Micro Press of Tokyo Institute of Technology

Fig. 3. Picture of Micro Punch and Fabricated Micro Hole

3 Development of Desktop Micro Forming System and Compressive Forming Die-Set

The size and load capacity of the desktop micro forming system must have a desktop size and to under 5Tonf as result that achieve investigation and research about optimal size and load capacity that equipment of commercialization concept must have as progressing research for press equipment's miniaturization to form micro component [1-2]. Forming system that has a specification more than this is classified into that it is suitable to product macro sized components that have milli size than micro. In this research, the desktop micro forming system planed to have target specification in Table 1. Structure and design optimization are achieved according to the target specification in Table 1 and can be obtained the final 3D CAD model such as Fig. 4. The desktop micro forming system is designed with precision actuating guide and flexible forming performance. The size of system is 260×340×655 (W×D×H, mm). A precision guide module consists of AC servo geared motor, gear module, precision ball

Table 1. Specification of Modified Desktop Micro Forming System

Specification	Description
Size	260×340×655 (W×D×H, mm)
Forming Capacity	4.0 Tonf
Max. Speed	400 mm/min
Stroke Resolution	0.1 μm
Actuating Type	AC Servo Geared Type

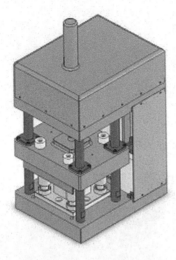

Fig. 4. CAD model of desktop forming system   **Fig. 5.** Manufactured desktop forming system

screw and guide 4-post module. The picture of manufactured desktop micro forming system with this optimal design result is displayed in Fig. 5.

This manufactured desktop micro forming system is designed and manufactured could be suitable to various micro components forming process. This system use a exclusive die set that can form the component between base part and cross head part of system, so that can apply this system to flexible micro forming application.

This desktop micro forming system is operated using AC geared servo motor and high precision ball screw, and applied a geared module as additional so that can regulate a forming speed and increase the forming load capacity. The gear module and AC servo motor structure's pictures applied to this system are shown in Fig. 6.

(a) Gear driven module   (b) AC servo motor is attached to gear driven structure

Fig. 6. Gear module and AC servo motor structures to make a variable forming speed

The desktop micro forming system and die set to manufacture a micro patterned component must have a micron size accuracy, so the precision displacement control fewer than sub-micro must be needed to prevent a damage by an over load of die set and system. The linear displacement sensor that have a resolution of 0.1μm is applied to the system to solve these precision displacement control problem. And a precision displacement feedback control fewer than sub-micro can be achieved using this sensor. The linear displacement sensor's picture that have a resolution of 0.1μm is shown in Fig. 7.

Fig. 7. The linear displacement sensor that have a resolution of 0.1μm

4 Manufacturing of Compressive Die Set for the Micro Herringbone Pattern

This research established by research target that achieve the micro herringbone pattern on the metal substrate. This micro herringbone pattern is used to generate the

    (a) Manufactured die set   (b) Equipped die set to the micro forming system

Fig. 8. Manufactured die set of a herringbone pattern forming and equipped die set module to this micro forming system

hydraulic dynamic pressure in so many application fields. The pictures of a manufactured die set of a herringbone pattern forming and equipped to this micro forming system are shown in Fig. 8.

5   Forming Results of Micro Herringbone Pattern on the Metal Substrate

In this research, the micro herringbone pattern is formed on the metal substrate (Stainless sintering substrate). The forming load of the micro herringbone pattern forming on the SUS sintering substrate is about 21kN. The precision machined mold is shown in Fig. 9 and forming results are shown in Fig. 10. The micro herringbone pattern is formed with about 10~20 microns height and good quality shape. Fig. 11 indicates that we can acquire the micro spiral pattern with the depth of 14μm above the forming load of 1300kgf, which is acceptable in the application of actual FDB in hard disk drive.

Fig. 9. Manufactured mold for forming the herringbone pattern on the metal substrate

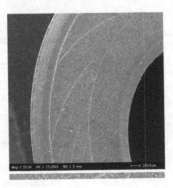

Fig. 10. Top-view SEM of FDB (forming load:1,400Kgf)

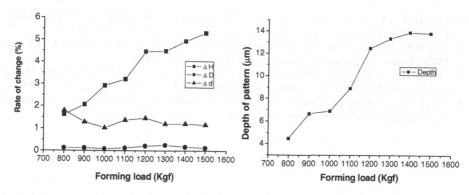

Fig. 11. Deformed shape of the herringbone pattern with various forming loads

6 Conclusion

This paper deals with a novel technique for the fabrication of spiral grooves in a dynamic thrust bearing. The main scheme proposed in this paper is to fabricate the microgrooves using desktop forming system. At first, a desktop press system was newly designed as shown in Fig. 1. The micro press system and die set to fabricate herringbone patterns should have a micron size accuracy, so the detailed displacement control fewer than sub-micro must be needed to prevent a damage by an overload of the die set and system. The linear displacement sensor that has a resolution of 0.1um is adopted to the system to guarantee the precise displacement control. Using this sensor, precise displacement feedback control fewer than sub-micro can be achieved. After that, die sets were also designed with the aid of finite element analysis. Tool dimension and shape of die sets shown in Fig.2 were determined from the FE results. Finally, micro forming tests were conducted with the developed desktop system and die sets. The testing material is sinter-forged cupper used in the actual FDB. In order to evaluate the formability and forming accuracy, SEM image of fabricated herringbone patterns were observed after the test. Experimental results demonstrate that herringbone grooves in FDB can be fabricated micro forming method using desktop press system to keep the product manufacturing time and costs low.

Acknowledgement

This work has been financially supported by Ministry of Knowledge Economy in Korea through Strategic Technology Development Project (Development of Micro Functional Precision Components Manufacturing Technology). The authors are also grateful to the colleagues for their essential contribution to this work.

References

1. Lee, H.-J., Lee, N.-K., Choi, S.: Development of Miniaturized Micro Metal Forming Manufacturing System. Materials Science Forum 544-545, 223–226 (2007)

2. Lee, H.-J., Lee, N.-K., Lee, S.-M., Lee, G.-A., Kim, S.-S.: Development of Micro Metal Forming Manufacturing System. Materials Science Forum 505-507, 19–24 (2006)
3. Joo, B.Y., Oh, S.I., Son, Y.K.: Forming of Micro Channels with Ultra Thin Metal Foils. CIRP Annals - Manufacturing Technology 53(1), 243–246 (2004)
4. Groche, P., Schneider, R.: Method for the Optimization of Forming Presses for the Manufacturing of Micro Parts. CIRP Annals - Manufacturing Technology 53(1), 281–284 (2004)
5. Tourki, Z., Zeghloul, A., Ferron, G.: Sheet metal forming simulations using a new model for orthotropic plasticity. Computational Materials Science 5, 255–262 (1996)

Miniaturization of Flexible Screwing Cell

Asser Vuola, Riku Heikkilä, Timo Prusi, Mikko Remes, Petri Rokka,
Niko Siltala, and Reijo Tuokko

Tampere University of Technology, Department of Production Engineering
Korkeakoulunkatu 6, 33720 Tampere, Finland
{asser.vuola,riku.heikkila,timo.prusi,mikko.remes,
petri.rokka,niko.siltala,reijo.tuokko}@tut.fi
http://www.tut.fi/tte

Abstract. The research in mini, micro and desktop factories originates
from the early 90's and has continued since then by developing the technological basis and different technological building bricks and applications in the field of high- precision manufacturing and assembly of future
miniaturised and micro products. This has paved the way for mini, micro
and desktop factories which are seen as one potential solution for what
kind of production by improving space, energy and material resource utilisation and answering to the needs of design for postponement and cus-
tomer-close customisation and personalisation. This paper presents one
case application for flexible micro factory. Application area is macro
world assembly system in miniaturised form. Current trend in this research is the miniaturisation of macro world machines and systems towards more sustainable production technologies.

Keywords: Desktop factory, micro factory, parallel robot.

1 Introduction

This paper describes the miniaturised concept for flexible screwing cell used in this
case for final assembly of a mobile phone. Flexibility has always been one key issue
in modern production research. In this application flexibility is defined as follows:

- Work piece fixture and material logistics: No product specific pallets, flat
  belt conveyor
- Material feeding: Ability to use screws of many sizes easily.
- Machine vision: No fixed sensors or assembly jigs
- Adaptive and easily re-configurable control system

When defining main functionality of assembly device or system, the most important functionality is how to handle and move parts and other materials. Thus we
can say that when ever discussed about the assembly system, robot or manipulator
must be defined. Typical macro world assembly cell contains a robot or manipulator of some type and its peripherals. When miniaturised assembly systems are designed or defined, all these aspects must be also covered. Miniaturisation of
this demonstrator bases on the following aspects [1] [2] [3]:

S. Ratchev (Ed.): IPAS 2010, IFIP AICT 315, pp. 309–316, 2010.

- Small, but fast and accurate enough screwing manipulator: In this case a belt driven parallel cartesian manipulator was selected
- Flexible and small feeding device for small parts
- Flexible work piece transportation (flat belt conveyor)
- Integration of the system in 30x20cm sized module.

2 Screwing Manipulator

Manipulator is the key issue in the assembly system and the most demanding part in the miniaturisation process. Conventional assembly manipulator is typically cartesian, SCARA- or arm-type robot. Common problem for all these types of mechanisms is the scalability of the moving mass relatively to the performance values. Many practical difficulties like cabling, mountings, friction, etc. also occur when one is trying to minimise these structures. Most critical performance deterioration is the decrease of the relative repeatability, caused by the hysteresis in the gear drives and bearings. The use of parallel kinematics in smaller scale robots has been seen as a promising method. Parallel robots are closed loop mechanisms presenting very good performances in terms of accuracy, velocity and rigidity compared to their own size and mass. In fig. 1 and 2 present some typical types of miniature robots using parallel kinematics. [4]

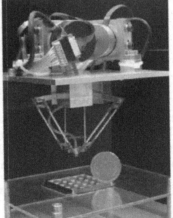

Fig. 1. PARVUS parallel scara-robot [4]   **Fig. 2.** Pocket delta robot

The main advantage of parallel kinematics in small scale devices is that the electric motor actuators are not a part of the moving mass. The mass of the actuators comes more dominant compared to the mass of the whole device when size goes smaller, especially if sufficient high accuracy is demanded.

So called "H-type" structure has clearly become more common in low-cost but still enough high performance industrial manipulators. Mechanism is belt driven parallel structure, where two motors move the same closed loop belt. Thus, motors do not correspond directly to cartesian X- and Y-directions directly. The working principle of the structure is seen in fig. 4. The overall structure of the manipulator is shown in fig. 3. This mechanism was selected for moving the screwing manipulator in XYplane.

Motors M1 and M2 have encoders at the end of the motor shaft. Position is therefore measured only from the driving motors. Indirect feedback and hysteresis caused by the belt drive are the main weaknesses of the structure and thus relatively poor repeatability of positioning was expected. In high precision systems, direct feedback and small hysteresis are the minimum requirements for better accuracy.

Kinematics of the device is quite simple although the motors do not correspond X and Y axes directly. X motion can be derived by moving both motors in the same direction and running the belt through the system. Y motion is respectively derived by running the motors in opposite directions. Idea of the X- and Y-motions is presented in fig. 4. Command values for driving the motors M1 and M2 are presented in equations 1 and 2. First term in both equations describes X-motion and second term Y-motion.

$$M1.x.y. \tag{1}$$

$$M2.x.y. \tag{2}$$

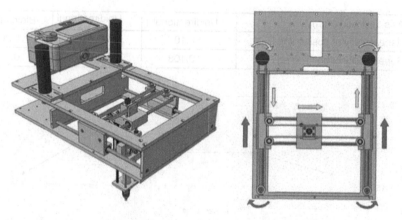

Fig. 3. CAD-model of the H-type structure   **Fig. 4.** X- and Y-motions

Z-axis has been implemented with a small ball screw. All guides are linear bushings due to their relative low price and availability compared to the very small prismatic guides. There is also small additional pneumatic extra stroke in Z-direction. With this small cylinder constant pressing force can be achieved for the screwing head by controlling the pressure in the cylinder. Actuator is Festo ADVC-6-10 two actions pneumatic cylinder with .6 mm piston and 10 mm stroke.

All motors in the manipulator are Maxon EC-motors, with planetary gearheads, hall sensors and optical encoders. Motion control drives are three small Maxon EPOS 24-1 motion controllers, which are connected together using CANOpen-bus. Position control is made in the embedded PC of the control system.

Screwing head itself is commercial hand held Atlas Copco MicroTorque ETD M 10 A. It is smallest commercially available electrical screwdriver. Driver has separate control box and it is freely programmable "intelligent" screwdriver. Communication is done with simple IO-handshaking. All process parameters are pre-programmed into the screwing controller. Several different screwing programs can be selected.

Accuracy Measurements

Repeatability of the positioning of the screwing manipulator was measured according to the ISO 23 0-2 standard [4]. Tests were carried out with Renishaw RLE20 2-axis plane mirror laser interferometer. Test setup is presented in fig. 6. Five target positions and two turning points with 16mm steps were used for X- and Y-axes, covering the whole range of the axes. As described in the standard, each target position must be measured five times in both directions. Positional deviation is calculated and collected for each point. Repeatability is then calculated from the mean positional deviations and standard deviations of the collected data. Repeatability of positioning R (equation 3) was calculated for both directions, and combined in bidirectional repeatability. The first results of the repeatability tests are shown in table 1.

Table 1. Repeatability of positioning

Axis deviation (mm)	Unidirectional ↑	Unidirectional ↓	Bidirectional
Repeatability of positioning R, X	0,107	0,028	0,201
Repeatability of positioning R, Y	0,108	0,045	0,165

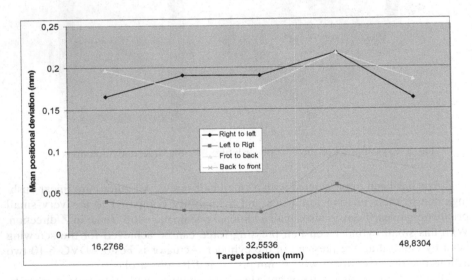

Fig. 5. Mean positional deviations

$$R_\cdot\max[R_i] \ . \tag{3}$$

Where R$_i$ indicates bidirectional repeatability of positioning at position i [5].

$$R_i = \max[2s_i \uparrow + 2s_i \downarrow + |B_i|; R_i \uparrow; R_i \downarrow] \ . \tag{4}$$

Where s$_i$ indicates the estimator of the unidirectional standard uncertainty of positioning at a position i [5].

As seen in fig. 5, unidirectional mean positional deviations for both axes are surprisingly good, but there is clearly around 0,15mm difference between the approach directions, which clearly indicates the hysteresis of the system.

4 Machine Vision

Machine vision is used in two purposes. Firstly to locate screws from the feeder and secondly to recognize work piece features. The Idea behind is that work piece position and orientation is not fixed. Using cameras and machine vision supports the idea of flexibility and easy re-configuration. Vision system is implemented with two small Eye machine vision cameras by IDS. Both cameras have standard C mount- objectives and USB-interface. The two cameras can be seen in fig. 6.

Fig. 6. Test setup and overall structure of the system

Feeder camera has objective with1 2,5 mm focal length and it gives 100 x 130 mm field of view (FOV). Conveyor camera has objective with 25 mm focal length and it gives 10x13 mm FOV. Whole working area of the manipulator can thus be covered. Conveyor belt and feeder plate is semi transparent and there is LED backlight inside the structures, but these are not used in this case. But front light is required in both cases.

Phone and screws are found by using simple template matching. From the phone the empty screw holes are a bit harder to find. When phone is black and the holes are black the holes cannot be found without special illumination. Because the holes are hard to find the already screwed holes are detected by using template matching. After the already screwed holes are detected, the places for empty holes are calculated relatively to the phones position and orientation. This is not as accurate as detecting the empty holes but this method does not rely on different illuminations.

5   Integration and Control

The presented assembly system has been built to support TUT-Microfatory concept. The main ideas in the concept are modularity and easy re-configurability. Modularity in this case means distributed control architecture and standardised communication and software interfaces. Whole screwing device should be considered as one additional process module with its own control system, which could be assembled to any other base module in the factory. Ethernet is used as higher level communication interface inside the factory concept. Schematic of the control architecture is presented in fig. 7 [6].

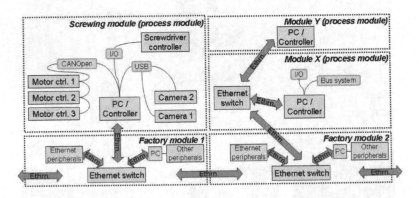

Fig. 7. Control system architecture between the modules

The module or cell internal control structure has more freedom. The used control architecture and bus system can be freely selected by the module provider, until the ethernet is available for communication with other modules. The detailed control architecture of the represented case is shown in Fig 8. A PLC with interpolated numerical control is the master of the whole system. It is programmed through the NC code. The controller is Beckhoff CX1020 with Windows XPe operating system. It

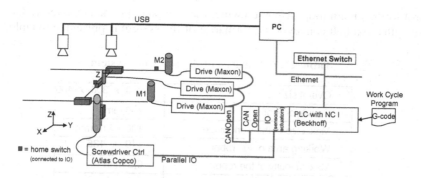

Fig. 8. Control system of the screwing cell

executes the drives for motors (Maxon EPOS 24/1) through CANOpen in velocity mode. Some parallel IOs are connected: Screwdriver controller, sensors like limit switches and actuators like lights or pneumatic valves. The controller communicates with the PC for image capturing and processing actions. PC has two USB cameras which are used to provide information for position and rotation correction before picking and assembly operations.

6 Conclusions

In the future the controller for TUT-Factory modules will be changed for smaller, because the industrial components are too large in size to fit into the module. Highly integrated mini PC-board or embedded controller is to be used. This will allow compacted size, optimised feature selection according to the needs, to name some. Current control setup was selected because of stability and reliability of the control, flexibility in IO modules, development time, focus in mechanical solutions and demonstrating the limitations of off-the-self industrial control components within the micro scale production modules.

Complete working cycles was successfully demonstrated after integration and calibration of the system. As described earlier in this paper, compact volume or accuracy were not the main goals for this demonstrator. Miniaturisation of macro world assembly system definitely sets some limits for the performance, but several advantages could be presented in terms of cost efficiency, energy and volume savings.

Certain level of flexibility can be reached by using machine vision as a feedback method for work piece and part localisation and flat belt conveyor for material logistics, instead of product specific pallets.

Complete system test was carried out with a real mobile phone case product containing six M1,6 x 3mm screws. Product is first identified and located with machine vision. Locations of the screws are then calculated in the frame of the manipulator based on a known design data of the case product. Systems then executed a loop until all screw holes are filled. Machine vision is used as feedback and quality control for this main loop. After screwing, the complete phone is moved to the next phase with a flat belt conveyor. Performance of the system is presented in table 2.

Table 2. Preformance of the screwing system

Overal size	200 x 300 x 400
Conveyor width	100 mm
Max. size of the work piece	100 x 110 mm
Working area of the robot	110 x 135 x 50
Max. velocity of the robot	0,5 m/s
Repeatability of the robot	0,2 mm
Possible screw sizes	M1 – M2,5 x 6mm
Max. Number of screw feeders	3
Average cycle time per screw	12 s

References

1. Heikkilä, R., Karjalainen, I., Uusitalo, J., Vuola, A., Tuokko, R.: The concept and first applications of the TUT Microfactory. In: IWMT 2007, 3rd International workshop on microfactory technology, Seogwipo KAL Hotel, Jeju-do, Korea, August 23-24, pp. 57–61 (2007)
2. Heikkilä, R.H., Karjalainen, I.T., Uusitalo, J.J., Vuola, A.S., Tuokko, R.O.: Possibilities of a Microfactory in the Assembly of Small Parts and Products - First Results of the M4-project. In: Proceedings of IEEE International Symposium on Assembly and Manufacturing, ISAM 2007, Ann Arbor, Michigan, USA, July 22-25, pp. 166–171 (2007)
3. Okazaki, Y., Mishima, N., Ashida, K.: Microfactory and Micro Machine Tools. In: Proc. of Korean-Japan Conference on Positioning Technology, Daejeon, Korea (2002)
4. Burisch, A., Wrege, J., Soetebier, S., Raatz, A., Hesselbach, J., Slatter, R.: "Parvus" a Micro-Parallel-Scara Robot for Desktop Assembly Lines. IFIP International Federation for Information Processing, vol. 198 (2006)
5. ISO 230-2 Test code for machine tools – Part 2: Determination of accuracy and repeatability of positioning numerically controlled axes
6. Siltala, N., Heikkilä, R., Tuokko, R.: Architectures and Interfaces for a Micro Factory Concept. In: Ratchev, S., Hauschild, M., (eds.): IPAS 2010, IFIP AICT 315, pp. 293–300 (2010)

Chapter 10

Micro-assembly Technology Studies

A Cooperation Model and Demand-Oriented ICT Infrastructure for SME Development and Production Networks in the Field of Microsystem Technology

Markus Dickerhof

Karlsruhe Institute of Technology
Institute for Applied Computer Science
Hermann v. Helmholtz Platz 1
D 76344 Eggenstein Leopoldshafen
Tel.: +49 7247 825754
Fax. +49 7247 825786
markus.dickerhof@kit.edu

Keywords: Collaborative engineering, new collaboration models Distributed and collaborative networks Readiness for participation in Virtual Enterprises.

1 Introduction

Like no other industry, until today microsystems technology is dominated by a few, large companies covering the whole development and production chain. Besides limitations coming from huge costs for equipment and manufacturing (especially in the field of Si-Microsystem technology) one of the reasons for the weak position of Small Medium Enterprises (SME) in this branch refers to organizational issues, arising from the specific surrounding conditions in this highly interdisciplinary and knowledge intensive field. Especially the smaller SME´s lack of sufficient human resources and an effective management of cross company knowledge about complex Microsystems development to cover all aspects of a complex and parallel product and process development as it is inherent in the one product- one process paradigm of MST.

Since most of the Microsystems technologies until today haven´t reached an almost mature state it near at hand, that there is a demand for approaches, offering the chance to effectively handle the riscs of a MST product development – which basically is not a technical but an organizational problem. This aspect gains more and more importance when the number of companies involved in such a process development increases. If SME networks seriously want to compete with larger companies, new approaches on how to handle a distributed knowledge intensive product development process need to be found.

2 A New Concept for Small Medium Enterprise Cooperations in the Field of Microsystems Technology

The starting point for bases on the perception, that neither the type nor the intensity of a cooperation can be seen as a static process. Much more the aspects mentioned before

S. Ratchev (Ed.): IPAS 2010, IFIP AICT 315, pp. 319–328, 2010.

depend on the intensity of the cooperation which was identified as a function of a chronological network development and of the intended purpose of the cooperation, both significantly influencing the choice of appropriate tools and methodologies.

Figure 1 gives a schematic overview of a classification scheme, allowing for the arrangement of organizations in four types of cooperation intensity. Each of these type has been identified to be existing in / of relevance for the Microsystems community. While the weakest type of cooperation, the market place (e.g. IVAM) primarily bases on the idea of a non-binding common platform for marketing, cooperation type four, the so called "Clan-Cooperation", represents the strongest one [Picot93]. A modified version of this approach has been identified as the most intensive and probably the most promising one for innovative MST companies in a German national funded project [MWF03].

The MST-"Clan-Cooperation" idea bases on the assumption, that in a MST cooperation the partners do not only have to share their production capacity in terms of what was called "virtual enterprise" in the late 90´s [VIRT09]. Furthermore the also share intensively their production and application knowledge enabling them to offer "one-stop-shop" services to their customers, e.g. as an OEM solution provider.

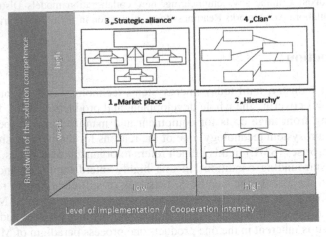

Fig. 1. SME-cooperation types in microsystems technology (modified model acc. to [PICO93])

Looking at the clan-cooperation model one can imagine that the inherent flexibility of this model provides significant advantages in the MST product development phase, but it is also clear, that the non-binding structure of the clan model finally is not the appropriate approach for a later production either in a small or in a medium or large scale. In this phase of the product development cycle the organizational framework certainly will result in a type of "general suppliership", conducted one of the partners out of the consortium. This general supplkier takes the lead in the consortium and integrated all other partner as subcontractors, probably using technologies introduced in the Virtual Enterprise context [MUEL99]. Hence, as soon as the MST product life cycle passes from the development to the production state, its differs no longer very much from other macro world production consortiums, as it has at least to adress the same requirements on quality, reliability, documentation, and productivity as other macro-scale components, integrated by large companies into their macro scale systems.

3 Demand-Oriented Software System Support for MST Development Process

The following perceptions have been made while a detailed analysis and discussion of the Clan cooperation model [DICK08]:

- Organisation and support tools must be flexible enough to support information sharing and the need for communication in an appropriate manner.
- In parallel to the rising amount of projects, and with the rising degree of integration of the network, the coordination should select appropriate tools in a demand-oriented manner. This means that in most cases not the most ambitious technological are the best solutions. Furthermore the ICT Infrastructure needs to be scalable and fittable. However the state of the art R&D projects in this field still focusses on the development of information and communication environments for mature organisation in the supply chain or large company context and do not yield appropriate solutions for SME networks.
- Information management, content management, shared workspace environments, videoconferencing can bring – due to the reduced amount of technical meetings - real value to all partners at a very early stage of a cooperation.

Based on these perceptions five basic ICT-requirements have been identified for the clan cooperation. A customized ICT infrastructure for a joint development and production in Microsystems technologies has

- to address the different phases of an application / production development.
- to support the heterogeneous infrastructures in the participating companies.
- to be flexible enough to deal with the different intensity of information exchange referring to the different phases of the cooperation life cycle. While the knowledge sharing intensity in a market place rests almost on the same level or raises up and decreases in a temporally limited cooperation like the virtual enterprise, it may continuously increase in a strategic alliance or a clan cooperation (Fig. 2).
- to address the specific constraints of the MST, resulting out of the still insufficient maturity of the technology itself. In most cases, the partner will not only have to deal with a cross company application development but also with a cross company production process development / adoption of established processes.
- to consider, that standardization in the information exchange on shop floor level (standardization of transport equipment) as well as on the logistic level is very weak in this community and in most cases belongs to the requirements of an end customer (if there exist one).

An effective and user centered infrastructure has to deal with all requirements mentioned above, if it shall not fail in terms of appropriateness and user friendliness.

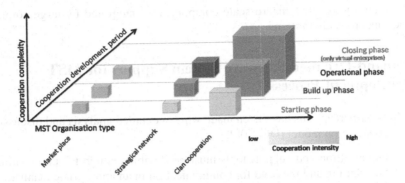

Fig. 2. Timely development of the cooperation intensity in typical SME networks related to the microsystems branche

4 Software Support for the MST Development Process on Organizational Level

In Section two came out, that for the production phase and from the organizational point of view a MST-SME cooperation acts similar to a macro production network, using almost the same tools for monitoring and control. Due to this work on MST specific software aspects will sooner or later focus on the technological aspects of assisting distributed teams while the knowledge intensive product development process. The following chapter summarizes a solution concept, supporting the "Open solution-paradigm".

Supporting collaborative teams through workflow technologies
Coming from the above mentioned "one product–one process paradigm" MST- SME cooperation -in contrary to "macro world"-production networks have to deal with a parallel process and product development which basically can be described as an "open-solution" process (Fig. 3).

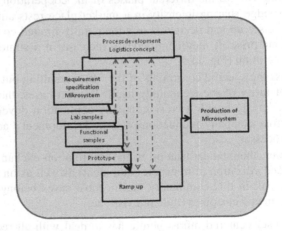

Fig. 3. Parallel process and product development in MST

"Open-Solution" process in this context means, that the final solution strategies for either the product- or the related production development process are not clear at an early phase of MST component developments.

This implies that the product development process itself can´t be finally specified from the beginning on. Other the hand side it does not mean that single steps in the development can´t be well specified, executed and analyzed which leads one to the conclusion that MST development processes need to be planned in a way that allows for a break down in small manageable pieces.

An appropriate solution path to handle such "solution open" developments is to describe the development steps indirectly through the definition of milestones for each phase that have to be achieved after a certain time or number of iterations. The milestones again can be clustered in such a way, that a set of milestones describe specific development phases, where at the end both, the customer and the project consortium, have to decide whether to continue or not. Anyway the solution-open problem still remains unsolved at this point.

For the modeling of this specific constraint we modified the Spiral-model, introduced by B. W. Boehm [BOEHM88]. The spiral model was introduced in 1988 and is until today one of the major software development models for handling the iterative software development process in an efficient way. We adopted the model to the needs of MST development, allowing us to describe each process by four basic sections, valid for all phases of the MST product development process: the brainstorming phase, the elaboration phase, the internal review and the external review (Fig. 4, Sections A-D). If the single development step was accomplished with satisfying results the development process continues to the next phase. If not, the same sequence has to be processed again.

In a next step we modeled each project phase as a sequence of milestones, clustered For MST collaborations a set of almost generic milestones had been elaborated, allowing for the above mentioned partition of the project in manageable slices (Fig. 5).

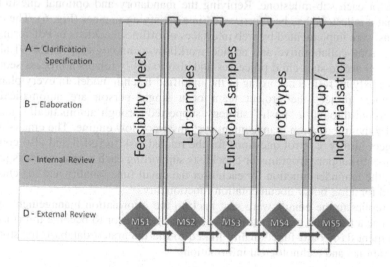

Fig. 4. Framework for the "solution open" MST- Product development

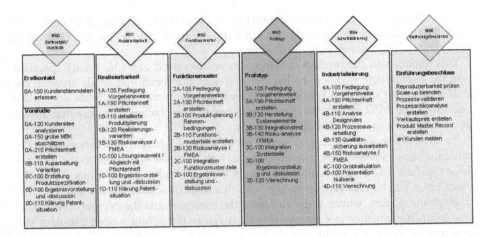

Fig. 5. MST- Design Control and Commercialization Guidelines of the German MicroWebFab Network

This set of milestones, called "Design, Control and Commercialization Guidelines" is organized in a way that supports the iterative processing of the product development.

Each sequence of milestones follows the same logic, introduced by the spiral model. Subtasks dedicated to each single sub-milestone fix the expected outcome of the development process (e.g. FMEA, Patent analysis, etc.). At detailed set of (generic) business process sequences ensures the comprehensive processing of the milestone -related activities. Templates for competences and responsibilities related to each process / milestone allow for a kind of standardized management of the product development process by coordination (Fig. 6).

To support the coordinator and the project team, a set of "control questions" was defined for each sub-milestone. Replying the mandatory and optional questions ensures that a milestone has been processed in a satisfying manner (Fig. 6). The control questions were implemented by web interfaces or offline checklists in Pdf-format.

Using semi-collaborative and ad-hoc workflow-technologies [SCHM99] allowed us to model and to start child processes and also to easily relaunch process sequences with unsatisfying results according to the modified Boehm model. In every phase of a development project the project team or a single person are automatically informed/asked through a "push-" strategy supported through automatically generated emails by the information management systems workflow engine. The emails include the project history, control questions for the related task to fulfill and also consist of hyperlinks to support programs or checklists supporting each single business process. Besides the reminder function for each user the email functionality and the checklists provided for a log-book/ documentation functionality.

The replies to the emails were sent back to the information management systems are used as a feedback mechanism allowing the coordinator to become informed via a management dashboard functionality linked to the integrated database for storage of the operational and technological information.

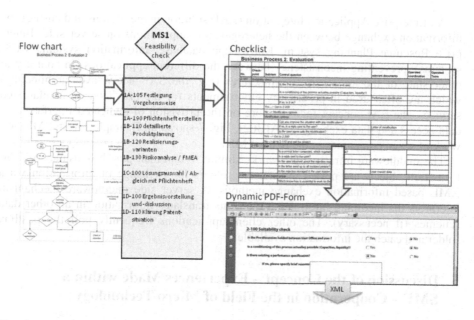

Fig. 6. Derivation of manageable information elements for the solution-open MST product development process

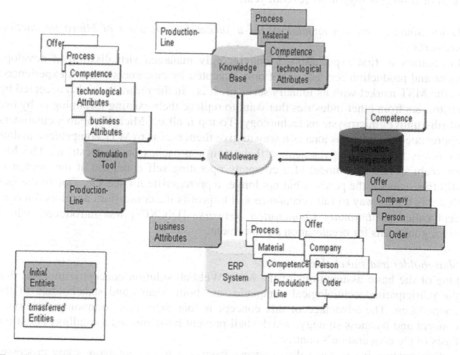

Fig. 7. Enterprise Application Integration Strategies to link SME software solutions in a demand oriented manner

An Enterprise Application Integration (EAI) solution in the background ensures the information exchange between the heterogeneous applications on server side (Enterprise Resource Planning system, Knowledge database, information system, MES-systems), preventing inconsistency between the different applications. In contrary to other state of the art service architectures (e.g. the SAP net weaver architecture) which are often used in proprietary environments for the provision of standardized services, the EAI-concept focuses more on the provision of dedicated interfaces for data access and data exchange [KAIB02]. Fig. 7 shows an Example for an SME network of seven companies in the field of MST.

The middleware infrastructure works similar to an email post box framework. Information objects are posted in different folders for exchange of information via a XML based information exchange methods. A server side middleware mechanism interprets the incoming information elements translates information in an other data schemes (if necessary). The other client applications frequently poll the postbox folder and catch the information available.

5 Discussion of the Concept – Experiences Made within a SME – Cooperation in the Field of Micro Technology

The following chapter summarizes the experiences having been made while the operation of the German MicroWebFab cooperation, a SME network operating in the field of Biotechnology in its seventh year.

Legal Bindings are not mandatory for a successful operation of Micro production networks
In contrary to first expectations a convincingly managed virtualisation of development and production services is not only accepted by customers that are experienced in the MST market with its foundry services. It is - in the meantime - also accepted by customers from other industries that want to replace their existing technologies by use of (disruptive) microsystems technology. To top it all off MicroWebFab´s customers appreciated the business idea as a way to give them access to MST competence and/or to assess MST alternatives against their own approaches on a shoestring. The MicroWebFab business model of a company operating self contained at the market is still the vision of the partners but not longer a prerequisite for success. Due to the fact that it is a long way to full acceptance and to profits the consortium developed a concept called *Evolutionary Cooperation Networks* [DICK08] was introduced, which gives guard rails for organisation advancement.

Shareholder and stakeholder structure
One of the basic assumptions of the MicroWebFab solution concept comprised that the participating technological companies are both, share- and stakeholders of the cooperation. The advantage of this concept is that there is a commonly developed concept and business strategy, which shall prevent basic misunderstandings about the basis of the cooperation's contract.

Furthermore it ensures a direct access from and to the company´s key processes and persons in the technological and organisational dimension. The proof of this basic thesis showed that the structure is one the one hand side a guarantee for proper implementation of the slogan "The sum is greater than its parts". This very strict assignment on the other hand produces certain sensitivity of the cooperation in terms of mutual dependence of partners.

Influence of the ICT infrastructure
The MicroWebFab example shows quite well, that a state of the art ICT Infrastructure with well defined interfaces on organisational and operational provides for some advantages in the competition with other networks or even larger companies. Effectiveness of communication and an extensive knowledge exchange can speed up the decision processes and with that lead to a kind of "first come, first serve"-effect. Finally internal effects such as the different trust level between the partners or the broad competence portfolio accessible to customers will be the primary drivers for the cooperation success.

6 Conclusion and Summary

More customer demand than ever are looking for one stop shop development and production services provided by SME cooperation. The necessary boundary conditions on organisational level can be also adressed by SMEs, as long as these are willing to intensively cooperate in a new quality of intensity and trust. An appropriate approach for this can be the clan cooperation. The modified MST clan cooperation concept of jointly a marketing and project management of equal SME partners is

- suitable for companies that are in the process of entering new markets
- suitable for smaller specialized companies with limited marketing/sales capacities,
- not suitable for companies with a focus on large scale /automotive market

Rigid and strictly structured IT-infrastructures are not an appropriate solution for this type of cooperation. Furthermore a demand oriented infrastructure has to be modular and flexible, where "flexibility" for an SME network in the context of MST has to address two topics: It has to address the specific requirements of a parallel product development and process adaption, which can be solved by a flexible process model based or an iterative development approach having been described in this paper. In addition it has to deal with restrictions in information exchange coming from the heterogeneous Information technology at the partner sites. To overcome these limitations, a demand oriented, EAI based approach supporting a cross company information exchange has been introduced.

References

[BOEHM88] Boehm, B.W.: A Spiral Model of Software Development and Enhancement. IEEE Computer 21, 61–72 (1988)
[DICK08]   Dickerhof, M.: Ein neues Konzept für das bedarfsgerechte Informations- du Wissensmanagement in Unternehmenskooperationen der Multimaterial Mikrosystemtechnik, Dissertation, Universitätsverlag Karlsruhe (2008)

[KAIB02]   Kaib, M.: Enterprise Application Integration. DUV wirtschaftsinformatik, Wiesbaden (2002)

[MWF03]   Gengenbach, U.: A project to establish a virtual enterprise for distributed development and fabrication of Microsystems. In: Proceedings of the COMS 2003 conference, Ypsilanti (2003)

[PICO93]   Picot, A., Scheer, A.W. (eds.): In Organisationsstrukturen der Wirtschaft und ihre Anforderungen an die Informations- und Kommunikationstechnik. Handbuch Informationsmanagement, Aufgaben - Konzepte – Praxislösungen. Gabler-Verlag (1993)

[SCHM99]   Schmidt, A., Düpmeier, C., Eggert, H.: WILDFLOW - components for building scientific and technical work environments. In: 1999 Internat. Conf. on Web-Based Modelling and Simulation, San Francisco, Calif., January 17-20 (1999)

[VIRT09]   N.N., Web portal of the California Network Virtual Enterprise, http://www.virtualenterprise.org/ (2009.07.20)

A Methodology for Evaluating the Technological Maturity of Micro and Nano Fabrication Processes

Emmanuel Brousseau[1], Richard Barton[2], Stefan Dimov[1], and Samuel Bigot[1]

[1] Manufacturing Engineering Centre, Cardiff University, Cardiff CF24 3AA, UK
[2] Cardiff Business School, Cardiff University, Cardiff CF10 3EU, UK
BrousseauE@cf.ac.uk

Abstract. Given that Micro and Nano Technologies (MNTs) is still an emerging field, it is important to adopt a tool for evaluating the maturity of MNT-based products and the production processes enabling their manufacture. In particular, as a risk assessment tool, it could help both the pace of technological adoption and the successful exploitation of these technologies. In this context, the objective of the research presented is to describe a methodology for assessing the maturity of MNTs. The paper also demonstrates the implementation of this methodology for a set of micro and nano manufacturing processes employed for tooling and replication. It could also be easily implemented to evaluate the maturity of other production processes for MNT-based products such as micro assembly technologies. The reported study was conducted in collaboration with two networks funded through the Sixth Framework Programme (FP6) of the European Commission (EC), namely the Multi-Material Micro Manufacture (4M) Network of Excellence (NoE) and the μSAPIENT Coordination Action (CA). By analysing data from R&D projects carried out in the field of MNTs by partner organisations in these two consortia, the maturity phases targeted by each project could be evaluated and as a result, the maturity profiles for given technologies could be extracted. An important output of this study is to help inform the industry, the global research community and policy makers about the current level of maturity reached by the MNTs which are developed in R&D projects carried out at European level and in particular, within the 4M NoE and the μSAPIENT CA.

Keywords: Technology maturity assessment, micro manufacturing, nano manufacturing, micro and nano technologies.

1 Introduction

Despite the recognised benefits that result from technology standardisation, whether directed at products or production processes, very little attention has been paid to propose methods for evaluating technological maturity consistently between organisations. Perhaps the most popular concept for performing such a maturity assessment is the Technology Readiness Level (TRL). This concept and the associated TRL scale were developed in the 1980s by the National Aeronautics and Space Administration (NASA) and further adopted in the 1990s by the United States Air Force [1]. It is a

S. Ratchev (Ed.): IPAS 2010, IFIP AICT 315, pp. 329–336, 2010.

measure to assess the maturity of an evolving technology such as materials, components or devices, prior to incorporating it into a system or subsystem.

Although the TRL evaluation method is well suited to the assessment of proprietary technologies developed by a single organization, it cannot be easily applied for obtaining a global picture of the maturity of technologies developed in parallel by different organisations as it is the case with MNTs. Thus, the objectives of this paper are to present an alternative methodology that would simplify the evaluation of technological maturity by different organisations and to apply this methodology in the context of MNTs. The motivation behind this research was also to obtain a picture at the European level of the distribution of the research efforts on MNTs along a technology maturity scale. Thus, it was anticipated that this study would help inform European and national funding bodies, the research community and the industry about the maturity of such technologies.

The paper is organised as follows. The next section presents the proposed methodology and describes its application in the context of MNTs. Then, the following section illustrates the maturity assessments results when this method has been applied to evaluate a set of micro and nano manufacturing processes belonging to the technological scope of the 4M NoE and the µSAPIENT CA. Finally, the generic findings of this study are presented to conclude the paper.

2 Methodology

The methodology employed in this study is illustrated in Fig. 1. It relies on identifying a portfolio of R&D projects in a given technological domain in order to have access to a rich and validated knowledge repository. The attractive characteristics of such an approach are that:

- The projects accessed are funded on a competitive base by regional, national and EC programmes that reflect specific industrial requirements and also the outcomes of roadmapping and foresight studies;
- The projects are also peer reviewed by experts that concur with the current status and targeted advances in key technology development areas that are stated in the project proposals;
- The projects involve consortia of industry and R&D partners that are specialists in their fields and have agreed a joint R&D programme.

In this research, a portfolio of MNTs-based R&D projects that involve partners in the EC FP6 funded 4M NoE and µSAPIENT CA was used to apply the proposed methodology. In total, these two networks bring together 40 R&D organisations spread over 17 different European member states. This portfolio comprised more than 300 projects which have commenced or have been completed over the last 5 years with European, national or institutional funding. The study is an attempt to position each of these research projects on a technology maturity scale in order to obtain a picture of the distribution of the MNTs R&D efforts across Europe. To carry out this analysis, five consecutive steps illustrated in Fig. 1 and described in the following sub-sections were identified.

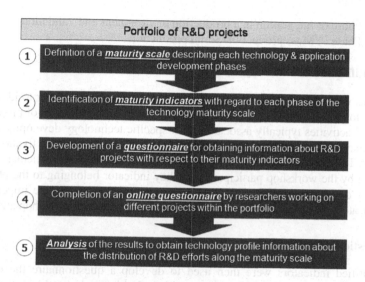

Fig. 1. Methodology for technology maturity assessment

2.1 Definition of a Technology Maturity Scale

The maturity scale used was inspired by the TRL scale which is composed of 9 levels grouped into 6 transition phases [1]. However, this scale had to be adapted to the context of this study as it did not describe appropriately the maturity levels of MNTs. For this reason, the TRL concept was presented to 4M and µSAPIENT partners in order to obtain their view on a maturity scale that could be applicable in the context of MNTs. This was done during a workshop that sought the input of thirty 4M and µSAPIENT experts. This workshop took place on the 14th February 2008 at the Fraunhofer Institut für Zuverlaessigkeit und Mikrointegration (IZM) in Munich. In particular, the participants were split into three groups, which were given the same questions throughout the workshop. However, each group was asked to provide answers in the context of particular R&D projects. More specifically, the first group had to give its input with respect to projects targeting the development of micro and nano manufacturing technologies. The second group concentrated on R&D projects that are focused on applying MNTs in different application areas. Finally, the remit of the third group was to provide input in the context of projects targeting both manufacturing technologies and application developments.

Each group was presented with the 6 transition phases of the TRL scale. Then, after an initial discussion they were asked to refine those phases taking into account the scope of the R&D work targeted and/or conducted in the context of their MNT projects. By combining the answers of the three groups, a common scale composed of the following seven "maturity phases" of technology development was identified:

- Phase 1: Basic technology research
- Phase 2: Feasibility study
- Phase 3: Technology development
- Phase 4: Technology demonstration
- Phase 5: System development/integration

- Phase 6: Integration in a production environment and validation
- Phase 7: Mass production/Serial production

2.2 Identification of Maturity Indicators

During this workshop, a Delphi-type study was also conducted in order to identify key indicators for each maturity phase. Thus, generic indicators in the form of project motivations and activities typically associated with specific technology development phases were obtained. For example, a key indicator identified as a possible motivation for setting up an R&D project was "new material to be developed". In particular, this was considered by the workshop participants as a key indicator belonging to the phase 1 of technology development: "Basic technology research". In addition, the importance of these indicators for each maturity phase was also weighted by the workshop attendees.

2.3 Questionnaire Design

The identified indicators were then used to develop a questionnaire the output of which would allow R&D projects to be positioned objectively on the maturity scale, without taking into account the application specific R&D issues addressed by them. In particular, its purpose was to present simple questions to researchers with respect to the motivations behind setting up an R&D project and the generic activities carried out within the project. At the same time, the answers to these questions were linked to the maturity indicators and thus, the maturity phases addressed by a project could be extracted automatically.

In order to collect the responses to the designed survey, a self-administered on-line questionnaire was preferred to other techniques such as semi-structured interviews for the following reasons:

- The identification of the maturity indicators during the workshop meant that the use of open-ended questions would be limited;
- The size of the targeted respondents was relatively large and geographically dispersed.

To design the questionnaire and come up with the necessary questions, a data requirement table was built as recommended in [2]. This ensured that the data collected would provide sufficient information to meet the aims of the survey.

2.4 Administration of the Questionnaire

For maximising the reliability of the responses, it was decided, whenever possible, to get the questionnaire completed directly by the individual researchers responsible for carrying out the work for each project considered in the survey. For this reason, the name and contact details of researchers associated with each project was obtained from the organisations taking part in the study.

In order to validate the questionnaire, a pilot survey was carried out with five researchers at the Cardiff University Manufacturing Engineering Centre. This sample was chosen because it represented a population with a similar profile to that expected to take part in the survey. As a result of this pilot study, a number of questions were modified in order to clarify their meaning and thus to avoid any confusion in their interpretation. Then, the final questionnaire was launched online on 13th June 2008 at the following address: http://www.surveys.cardiff.ac.uk/maturitymnt.

2.5 Analysis of the Results

In order to evaluate the maturity of the R&D projects from the data collected with the questionnaire, the use of maturity profiles was preferred to the calculation of single maturity value as it would be the case with the TRL concept. More specifically, a technology assessed with TRL is considered to have reached one particular level along the TRL scale. Instead, in the context of this study, a profile was chosen to represent the results of the maturity assessment because it provides a more realistic "snap shot" of current status of MNTs than that obtained with a single maturity value. For example, the micro milling technology is currently being exploited commercially by mould and watch making industries. At the same time, the research community recognises that further fundamental investigations are also needed to understand and especially to model the mechanics of mechanical machining at the micro scale [3]. In this case, the output of the maturity assessment method employed should result in a profile capturing the fact that micro milling is a technology for which R&D efforts span a broad spectrum of maturity phases.

Based on the responses received from the questionnaire, a maturity profile was constructed for each project. For example, one of the questions asked the respondents to select the main motivations/triggers to start an R&D project from the list of identified maturity indicators. If a given survey participant selected "new material to be developed", which was classified as a phase 1 indicator (see section 2.2), then based on the weight of this indicator, a particular score would be assigned for phase 1 for this project. Any other phase 1 indicators selected by this survey respondent would increase the score for this phase. Another question consisted in asking what were the essential tasks required to achieve the project objectives, again from the list of identified indicators. In this case, the same scoring procedure was applied to extract the scores for each maturity phase. Finally, during the analysis of the survey results for a particular project, the total score achieved for each phase was divided by the sum of the scores for all phases. This allowed expressing the score obtained for each phase as a percentage of that for all phases.

Thus, the maturity associated with each project could be presented as a profile displaying percentages of R&D efforts along the adopted maturity scale. In particular, the x axis of a graph displaying such a maturity profile represents the considered seven phases while its y axis indicates the percentage of R&D efforts in terms of technological motivations and research activities that are associated with each phase.

3 Maturity Assessment Results for Selected Micro Tooling and Replication Processes

The questionnaire was designed to differentiate whether a given manufacturing technology could be categorised as "developed" within a project or simply "used as a supporting technology". In the first case, a project would typically focus on overcoming one or several limitations of a manufacturing process in order to improve or broaden its capabilities. In the second case, a project would be more likely to develop a product incorporating micro and nano features and utilise one or a set of micro/nano manufacturing processes to produce different components of the developed product.

The maturity profile for a given manufacturing technology was calculated by grouping together all the projects which categorised it as "developed". The projects in which it was simply "used as a supporting technology" were not taken into account in this assessment because the maturity profiles of those projects would be representative of the particular application developed and thus less meaningful for evaluating the maturity of the associated manufacturing technologies. Based on the grouping of projects targeting the development of a given technology and the assessment of their respective maturity profile, the average percentage corresponding to each maturity phase was calculated. For example, in the case of two projects aiming at developing Nano Imprint Lithography (NIL), if the maturity profile for phase 1 had reached 30% for project A and 50% for project B, then the average percentage corresponding to phase 1 would be 40% for this technology.

As it was mentioned earlier, the use of maturity profiles was preferred in this study as a means to increase the information content of the maturity assessment results. However, in order to provide a maturity ranking and to compare one technology with another, it is also useful to complement the information given by the maturity profiles by deriving from them a single maturity value for each technology. To achieve this, the sum of the percentages for phases 5, 6 and 7 was computed for each technology. This sum is called the "maturity indicator" as it gives the percentage of R&D efforts that are targeted at the most mature phases of the scale and as a result, it provides an indication of the suitability of a technology to be exploited commercially.

Fig. 2 and Fig. 3 show the maturity profiles obtained for a restricted number of micro tooling and replication processes. These profiles are compared against each other

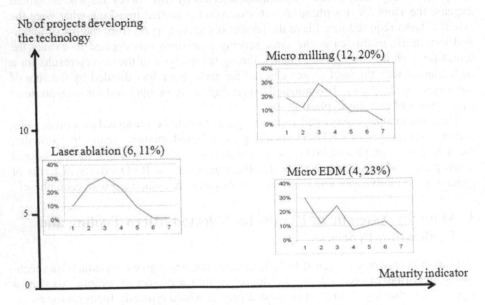

Fig. 2. Comparison of maturity profiles for micro tooling processes

by mapping them on a graph for which the x axis represents the value of the maturity indicator for a given technology while the y axis shows the number of projects surveyed that developed a particular technology. The processes shown in Fig. 2 and 3 correspond to those that were considered the most important for the future according to the results of a roadmapping study carried out in 2006 by the 4M community [4]. When considering micro tooling processes only, Fig. 2 shows that the maturity of micro milling and micro EDM is ranked higher than that for laser ablation. This fact tends to support the real impact of these technologies in the context of the micro tool making industry. Also it is not surprising to observe from Fig. 3 that, among the polymer replication processes, the maturity indicator of injection moulding is higher than that of NIL and nano imprinting.

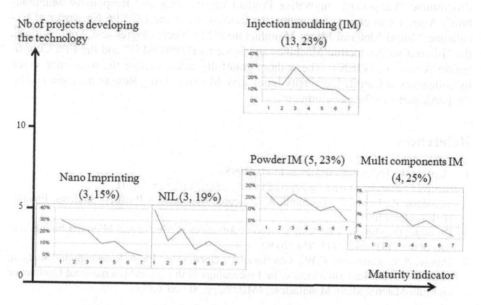

Fig. 3. Comparison of maturity profiles for replication processes

4 Conclusions

The paper presented a methodology for assessing technology maturity which is inspired by the TRL concept. However, it is designed to overcome some of the limitations of this concept. In particular, the proposed methodology was developed to simplify the maturity evaluation procedure in order to combine a large number of inputs from a rich and validated knowledge repository in the form of an R&D project portfolio. In addition, the method results in a maturity assessment output containing an increased information content and it also allows the identification of a broad picture of technology maturity that is not specific to a particular organisation.

The methodology was demonstrated for a set of micro and nano manufacturing processes employed for tooling and replication. However, it could also be easily applied to evaluate the maturity of a range of micro assembly technologies. The generic finding resulting from implementing this methodology on selected micro and nano fabrication processes is that R&D efforts exist to support the integration into production environments of the most important future manufacturing technologies, as identified by the 4M community. However, the proportion of these R&D activities is lower compared with those focused on the less mature phases of their development.

In addition to being a risk assessment tool for industry, the proposed methodology should be valuable for funding bodies and policy makers to monitor the impact of sponsored R&D projects and to identify funding gaps along the maturity scale.

Acknowledgments. The research reported in this paper is funded by the Engineering and Physical Sciences Research Council (GR/S75505, EP/F056745/1) and the ERDF Programme "Supporting Innovative Product Engineering and Responsive Manufacture". Also, it was carried out within the framework of the EC FP6 Networks of Excellence, "Multi-Material Micro Manufacture (4M): Technologies and Applications", the "Innovative Production Machines and Systems (I*PROMS)" and the FP6 Coordination Action μSAPIENT. The authors gratefully acknowledge the assistance given by colleagues at Cardiff University Innovative Manufacturing Research Centre and by the participants to the questionnaire.

References

1. Mankins, J.: Technology readiness levels (1995),
   http://www.hq.nasa.gov/office/codeq/trl/trl.pdf
2. Saunders, M., Lewis, P., Thornhill, A.: Research Methods for Business Students. Prentice Hall, Harlow (2003)
3. Dornfeld, D., Min, S., Takeuchi, Y.: Recent Advances in Mechanical Micromachining. Annals of the CIRP 55(2), 745–768 (2006)
4. Dimov, S.S., Matthews, C.W., Glanfield, A., Dorrington, P.: A Roadmapping Study in Multi-Material Micro Manufacture. In: Proceedings of the Second International Conference on Multi-Material Micro Manufacture, 4M2006, pp. xi–xxv (2006)

Function and Length Scale Integration Study in Emerging MST-Based Products

Samuel Bigot, Stefan Dimov, and Roussi Minev

The Manufacturing Engineering Centre
Cardiff University, Queen's Buildings, The Parade, Newport Road
Cardiff CF24 3AA, United Kingdom

Abstract. This paper discusses the issues typically occurring when integrating, into a single product, functions that can only be realised by employing different length scale features (macro, meso, micro and down to nano). Following the responses of a survey performed on European research projects, a trend for replacing assembly steps with multi-scale processing of single components is highlighted. Two main issues emerging from the implementation of such approach are then discussed. The first one relates to the concurrent design of materials properties and manufacturing processes. The second focuses on the need for new production approaches and organisational models applied by companies.

Keywords: Multiple-scale, functional integration, length scale integration, material design, organisation models, micro and nano manufacturing.

1 Introduction

New innovative products rely more and more on the integration of multiple functions in as small as possible enclosures, and thus on miniaturization that simultaneously broaden their functionality. Such products encapsulate in a single container [1] various functions emerging from different research field, for instance nanoelectronics, microsensors, micro and/or nano actuators or microfluidic, and would generally incorporate different length scale features, from the millimeter to the nanometer range. The concerted efforts of designers and manufacturing specialists to achieve such Function and Length Scale Integration (FLSI) can be justified by the numerous advantages that the adoption of this product development philosophy offers as reductions in cost, size, material usage and power consumption.

The FLSI challenge, which can be shortly defined as the integration of a range of functional features with different length scales into a single product, generally relies on emerging micro-assembly technologies. Alternatively, due to the intrinsic relationship between functional features and machined structures, the incorporation of various functions can sometimes be performed directly, without assembly, in a single component using emerging micro and nano structuring technologies. In any case, this is a difficult task due to the necessity to perform such integration reliably and cost effectively, at nano through micro to meso meter scales.

S. Ratchev (Ed.): IPAS 2010, IFIP AICT 315, pp. 337–342, 2010.

There are already many emerging product ideas and concepts based on FLSI, which are often a result of multidisciplinary R&D programmes. For instance, the following products are good examples of innovative applications of this design and manufacturing approach: (i) the polymer-based lab-on-chip platform (Figure 1.a) for protein detection in point-of-care applications developed in the European FP6 project SEMOFS [2], which aimed at integrating active optical components like a planar surface plasmon resonance sensors with a microfluidic system (including actuators); (ii) a contact lens encapsulating micron-scale metal interconnects in a biocompatible polymer (Figure 1.b) that includes light emitting diodes [3]. The described third generation contact lens is opening the door to a wide range of exiting new consumer products, such as a see-through display that could be both remotely powered and controlled via a wireless link with potential applications in gaming, training, and manufacturing [4].

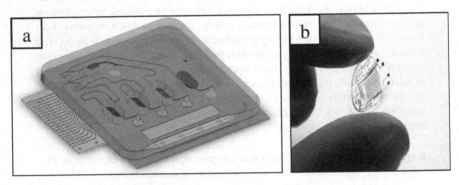

Fig. 1. a) Low-cost lab-on-chip b) Multifunction contact lens [3]

A common aspect in all these products is that the integration of various functions generally requires the latest state-of-the-art technologies, and in many cases the development of new manufacturing methods and chains of processes for incorporating different length-scale structures/features in a single component or the use of novel assembly techniques. The structure of the paper is as follows.

Based on a survey, the first section of the paper discusses the main integration methods used during the development of new MST based products. A trend for replacing assembly steps with multi-scale processing of single components is highlighted and the two main issues emerging from the implementation of such approach are then discussed in the last two sections.

2 Integration: From Assembly to Single Component Processing

A survey of 88 European research and development projects was conducted. All projects were focused on the development of MNT-based products and micro/nano manufacturing technologies that underpin their production while 54 of them involved companies as partners. One of the main objectives of this survey was to identify typical issues occurring when designing and manufacturing products incorporating functional features with different length scales. In this context, it was reported that 75 of the surveyed projects targeted the production of micro/nano components, while for 48 of them it was stated that the integration of micro and nano structures represented a significant problem.

The importance of integration using an assembly stage in micro and nano manufacturing was clearly supported by the results of the conducted survey where for 52% of the projects an assembly step was used in some way as part of the FLSI process, and it was the only FLSI process used in approximately 31% of the projects. Assembly is indeed a key stage in the fabrication of a product, which may require the use of various processes to integrate components manufactured separately due to different technical reasons, e.g. complexity, different material properties, capabilities restrictions and/or incompatibility of the available manufacturing processes [5]. 12 of the 44 FLSI issues reported were related to an assembly step. Particular concerns were the cost and reliability of bounding techniques used when dealing with micro components, and also dimensional accuracy of the components to be assembled.

For many economical and technical reasons and taking into account general rules from Design for Manufacture [6] it is generally better to achieve the required functionality of component or products using the minimum number of parts and thus to minimise the assembly steps in achieving FLSI in products, if not possible to eliminate them all together. This generic Design for Manufacture consideration was confirmed in the survey where 40% of the projects with as objective the development of new products use assembly as the sole procedure for FLSI while this is the case only with 15% of the projects targeting improvements of existing products. For such projects in most cases the improvements are sought / achieved through the use of a chain of multiple processes working on a single component, which suggests that this solution is preferred in the attempts to eliminate and/or minimise the assembly operations. By applying this approach towards FLSI, features with different length scales are machined directly on a single component. To achieve this, a sequence of compatible and complementary multidisciplinary processes are used, most of which can be classified as removal, deposition, deformation or replication ones. However, in implementing this approach new integration issues arise. Materials should be compatible with the set of micro/nano manufacturing technologies used in a process chain and specific organisational approaches are likely to be required to face the interdisciplinary issues and high capital investments linked with micro manufacturing.

3 Concurrent Material and Process Design

When integrating different length scale features, the microstructure of the materials that will undergo structuring should be optimised taking into account their specific machining response to various meso/micro/nano manufacturing technologies used in combinations while meeting the end product property requirements. This can be difficult to achieve as these requirements might differ significantly or even contradict each other.

Evidences of the need for the development of specifically tailored materials for micro-manufacturing are emerging, such as in the R&D programme on "Integrated Development of Materials and Processing Technology for High Precision Components" funded by NEDO in Japan [7] and the findings of the "International Assessment of Research and Development in Micro-manufacturing" in the USA [8].

There are a number of material deposition and refinement technologies (solid-solid, liquid-solid, vapour-solid, etc.), that can be readily applied to design/tune the material microstructure, and thus achieve a step change in their machining response. Examples of such technologies are Physical Vapour Deposition (PVD), Chemical Vapour Deposition (CVD), melt quenching, Electrochemical Deposition (ECD) or severe plastic deformation (SPD), with which it is possible to produce 'nano metals' that have more appropriate amorphous (glassy) structure or grain sizes down to sub-100 nm compared to a typical grain size of 100 μm in traditional metals.

Two recently completed feasibility studies illustrate the potential of such integrated approach for material and process design [9,10]. In the first study, by designing/optimising a Ni workpieces microstructure (e.g. grain size which controls the surface grain anisotropy) and the Focus Ion Beam (FIB) milling technology (ion fluence which controls the process dynamics) simultaneously it was possible to identify processing and performance windows that "overlap", and thus lead to synergetic effects on achievable feature resolution and surface integrity. The second study involved micro milling of two Al alloy workpieces that were metallurgically and mechanically modified, respectively. The result showed that by refining the material microstructure through SPD it is possible to achieve drastic improvements, more than three times, of surface roughness of thin features, 20 μm ribs, in micro-components [10]. In both cases, the synergetic material and process development led to a "step change" in surface quality, dimensional accuracy and functional properties of the final product.

Ultimately, to properly implement a concurrent product, material and process design, it is necessary, together with the process design, to optimise the micro-structure of materials and thus to improve their processing response to one or even a set of component technologies in a process chain (Figure 13). This could be achieved by refining/tuning the bulk material to achieve a significantly better machining response with the majority of the manufacturing processes and/or by locally "tuning" the surface layer, with sub-mm thickness, which will undergo structuring with other complementary manufacturing processes.

Using such material design approach for FLSI gives almost limitless structuring possibilities. The main issues are the production cost which can increase significantly due to the complexity and wide range of capital-intensive state-of-the-art technologies involved. However, with an appropriate business organisation, discussed in next section, such approach would be particularly applicable for the fabrication of replication tools, e.g. for injection moulding, glass moulding, hot embossing, roll to roll imprinting, stamping and coining, with which very large batches of parts can be produced, more than 100 000 per batch, and thus making the implementation of this approach cost effective.

4   New Production Approaches

Many SME's position themselves as partners of large companies and cooperate with them in a made-to-order fashion. Such a model gives little room, if any, for FLSI issues to be properly taken into account in the product design and manufacture stages, because they would necessitate the manufacturers to dynamically and cost-effectively integrate, optimise, configure, simulate, restructure and control not only their available in house surface structuring systems but also their materials' supply networks and part designs, in a coordinated manner. There are attempts to address these difficulties, e.g. in the production of micro parts by micro injection moulding. In particular, some replication machine providers work very closely with tool makers to develop and sell "packages", replication machines together with the necessary tooling and even to optimise processing parameters to satisfy specific product requirements, rather than focusing on each component technology separately. This approach could be broadened to include the material providers in the development of such solutions, and thus to offer materials optimised for specific FLSI requirements.

To deal with the interdisciplinary issues related to FLSI and the high capital investments required, new organisational approaches such as distributed production systems have to be considered. This will allow companies to design and produce parts without actually having all the required equipment and expertise physically on site, avoiding costly capital investments and personnel training. This is reflected in the responses obtained for the surveyed projects, where the majority of project consortia took the form of value networks or horizontal value chains, especially when considering projects focusing on the development of new products. Distributed production is not new in itself. For example, the dental industry uses it for the manufacturing of crowns, while the tool industry for the manufacturing of drills. But, in all existing cases, only variations of predetermined shapes can be produced. In a FLSI distributed system any complex 3D geometry at meso, micro and nano scale would have to be achievable. Designing and producing such products would require a highly integrated organisational model, bringing together a wide range of expert organisations.

Models such as extended enterprises and more specifically virtual enterprises, which can be viewed as a temporary alliance of companies taking advantage of a market opportunity [11], would have to be considered. Such approach will allow companies to focus on their key competencies without having to realise the complete process chain in house.

5 Conclusion

The results of a survey performed on a portfolio of 88 European research projects highlighted a inclination to replace assembly steps with multi-scale processing of single components when design MNT-based products. According to the survey the preferred methods for addressing the FLSI issues when developing new products are still the assembly ones. Nevertheless a trend for replacing them with multi-scale processing solutions can be identified, especially when improving MNT-based products. The challenges associated with the implementation of such solutions were discussed, too. In particular, the need for a new concurrent process-material development approach was presented in order to take full advantage of machining capabilities offered by component technologies in process chains, and thus to create the necessary prerequisites for achieving FLSI. Finally, the necessity of new organisational approaches was discussed to address technical and investments challenges that have to be tackled in achieving FLSI, e.g. finding different innovative ways for implementing distributed production systems in the MNT context.

Acknowledgments. The research reported in this paper is funded by the Engineering and Physical Sciences Research Council (GR/S75505) and the ERDF Programme "Supporting Innovative Product Engineering and Responsive Manufacture". Also, it was carried out within the framework of the EC FP6 Networks of Excellence, "Multi-Material Micro Manufacture (4M): Technologies and Applications" and the "Innovative Production Machines and Systems (I*PROMS)".

References

1. Evoy, S., DiLello, N., Deshpande, V., Narayanan, A., Liu, H., Riegelman, M., Martin, B.R., Hailer, B., Bradley, J.-C., Weiss, W., Mayer, T.S., Gogotsi, Y., Bau, H.H., Mallouk, T.E., Raman, S.: Dielectrophoretic assembly and integration of nanowire devices with functional CMOS operating circuitry. Microelectronic Engineering 75, 31–42 (2004)
2. Nestler, J., Morschhauser, A., Hiller, K., Otto, T., Bigot, S., Auerswald, J., Knapp, H.F., Gavillet, J., Gessner, T.: Polymer Lab-on-Chip systems with integrated electrochemical pumps suitable for large scale fabrication. International Journal of Advanced Manufacturing Technologies (2009)
3. Ho, H., Saeedi, E., Kim, S.S., Shen, T.T., Parviz, B.A.: Contact lens with integrated inorganic semiconductor devices. In: Proceedings of the IEEE 21st International Conference on Micro Electro Mechanical Systems, pp. 403–406 (2008)
4. Parviz, B.A., Lingley, A.: Multipurpose integrated active contact lenses. SPIE Newsroom (2009), http://spie.org/x35114.xml?ArticleID=x35114
5. Hansen, H.N., Tosello, G., Gegeckaite, A., Arentoft, M., Marin, L.: Classification of assembly techniques for micro products. In: Proceedings of the 1st International Conference on Multi Material Micro Manufacture, Karlshrue, Germany, pp. 283–286 (2005)
6. Boothroyd, G., Dewhurst, P., Knight, W.A.: Product Design for Manufacture and Assembly. CRC Press, Boca Raton (2001)
7. Hayashi, K., Kiuchi, M.: Integrated Development of Materials and Processing Technology for High Precision Components. Sosei Kako Shinpojiumu 243, 1–11 (2005)
8. WTEC Panel: International Assessment of Research and Development in Micro manufacturing (2005), http://www.wtec.org/micromfg/
9. Li, W., Minev, R., Dimov, S., Lalev, G.: Patterning of Amorphous and Polycrystalline Ni78B14Si8 with a Focused Ion Beam. Applied Surface Science 253(12), 5404–5410 (2007)
10. Popov, K.B., Dimov, S.S., Pham, D.T., Minev, R.M., Rosochowski, A., Olejnik, L.: Micromilling: material microstructure effects. Proceedings of the Institution of Mechanical Engineers-Part B 220(11), 1807–1813 (2006)
11. Szegheo, O., Petersen, S.A.: Extended Enterprise Engineering - A Model-Based Framework. Concurrent Engineering 8(1), 32–39 (2000)

Author Index